Sensors and Actuators: Technology and Applications

Sensors and Actuators: Technology and Applications

Edited by
Princeton Brown

Larsen & Keller
www.larsen-keller.com

Sensors and Actuators: Technology and Applications
Edited by Princeton Brown
ISBN: 978-1-63549-256-9 (Hardback)

© 2017 Larsen & Keller

Larsen & Keller

Published by Larsen and Keller Education,
5 Penn Plaza,
19th Floor,
New York, NY 10001, USA

Cataloging-in-Publication Data

Sensors and actuators : technology and applications / edited by Princeton Brown.
    p. cm.
Includes bibliographical references and index.
ISBN 978-1-63549-256-9
1. Detectors. 2. Actuators. 3. Automatic control. 4. Remote sensing. 5. Radar.
I. Brown, Princeton.
TK7872.D48 S46 2017
681.2--dc23

The publisher's policy is to use permanent paper from mills that operate a sustainable forestry policy. Furthermore, the publisher ensures that the text paper and cover boards used have met acceptable environmental accreditation standards.

Printed and bound in the United States of America.

For more information regarding Larsen and Keller Education and its products, please visit the publisher's website www.larsen-keller.com

Table of Contents

Preface   **VII**

Chapter 1   **Introduction to Sensors and Actuators**   **1**
            a. Sensor   1
            b. Actuator   4

Chapter 2   **Technologies Related to Sensors**   **9**
            a. Data Logger   9
            b. Metal Detector   14
            c. Photoelectric Sensor   22
            d. Global Positioning System   24
            e. Wireless Sensor Network   52
            f. Sonar   58
            g. Echo Sounding   77
            h. Level Sensor   80
            i. Biosensor   89
            j. Blood Glucose Monitoring   100
            k. Load Cell   104

Chapter 3   **Varied Types of Actuators**   **111**
            a. Pneumatic Actuator   111
            b. Hydraulic Cylinder   113
            c. Linear Actuator   126
            d. Plasma Actuator   145
            e. Rotary Actuator   150

Chapter 4   **Actuators: Technologies and Devices**   **155**
            a. Pneumatic Motor   155
            b. Pneumatic Cylinder   160
            c. Hydraulic Press   166
            d. Jackscrew   168
            e. Hoist (Device)   171
            f. Electroactive Polymers   174
            g. Roller Screw   182
            h. MEMS Magnetic Actuator   188

Chapter 5   **Remote Sensing: An Overview**   **195**
            a. Water Remote Sensing   195
            b. Remote Sensing   197
            c. Lidar   206
            d. ERDAS Imagine   219
            e. TerrSet   222
            f. Remote Sensing (Archaeology)   223

Chapter 6 **Radar and its Applications**   **228**
    a. Radar   228
    b. Radar Imaging   250
    c. Radar Navigation   255

Permissions

Index

Preface

The aim of this textbook is to provide in-depth information about the various applications and uses of sensors and actuators. It is designed in such a way that it provides the readers thorough insights about this subject. Sensors and actuators are combined to create an interface which is used for technological solutions. This textbook seeks to enumerate the technology and the future advances that are related to sensors to actuators. It mainly provides functional safety in machinery and emergency stop applications. This book is a compilation of chapters that discuss the most vital concepts in the field of sensors and actuators. It presents this complex subject in the most comprehensible and easy to understand language. While understanding the long-term perspectives of the topics, the text makes an effort in highlighting their impact as a modern tool for the growth of the discipline. For all those who are interested in this field, this textbook can prove to be an essential guide.

To facilitate a deeper understanding of the contents of this book a short introduction of every chapter is written below:

Chapter 1- Sensors are used to detect fluctuations in the environment and then to provide an output. Actuators are devices that provide mechanical movement on command. They are a part of our everyday life. This chapter is an overview of the subject matter incorporating all the major aspects of sensors and actuators.

Chapter 2- Sensors are used in everyday objects. Some of the technologies related to sensors are explained in this chapter. Data logger, metal detectors, global positioning system, sonar, echo sounding and load cell are all technologies related to sensors. The topics discussed in the chapter are of great importance to broaden the existing knowledge on sensors.

Chapter 3- Actuators can best be understood in confluence with the major topics listed in the following chapter. Actuators are of five types, hydraulic, pneumatic, electoral, thermal and mechanical actuators. Actuators in machines are responsible for moving or controlling the system. The aspects elucidated in this chapter are of vital importance, and provide a better understanding of actuators.

Chapter 4- The chapter serves as a source to understand the technologies and devices of actuators. Some of the technologies and devices involved in actuators are hydraulic press, jackscrew, roller screw, and pneumatic cylinder. Tools and techniques are an important component of any field of study. The following chapter elucidates the various tools and techniques that are related to actuator technology.

Chapter 5- Remote sensing is the receiving of information about an object with no physical contact. The military, intelligence and economic planners usually practice it. This chapter is an overview of the subject matter incorporating all the major aspects of remote sensing. This chapter is a compilation of various ideas of remote sensing that form an integral part of the broader subject matter.

Chapter 6- Countries use radar to detect aircrafts, missiles, vehicles and spacecrafts. It is a system that uses radio waves to determine the objects in the area concerned. It has proven to be of great importance to the military complex. The diverse applications of radars in the current scenario have been thoroughly discussed in this chapter.

I owe the completion of this book to the never-ending support of my family, who supported me throughout the project.

Editor

Introduction to Sensors and Actuators

Sensors are used to detect fluctuations in the environment and then to provide an output. Actuators are devices that provide mechanical movement on command. They are a part of our everyday life. This chapter is an overview of the subject matter incorporating all the major aspects of sensors and actuators.

Sensor

In the broadest definition, a sensor is an object whose purpose is to detect events or changes in its environment, and then provide a corresponding output. A sensor is a type of transducer; sensors may provide various types of output, but typically use electrical or optical signals. For example, a thermocouple generates a known voltage (the output) in response to its temperature (the environment). A mercury-in-glass thermometer, similarly, converts measured temperature into expansion and contraction of a liquid, which can be read on a calibrated glass tube.

Sensors are used in everyday objects such as touch-sensitive elevator buttons (tactile sensor) and lamps which dim or brighten by touching the base, besides innumerable applications of which most people are never aware. With advances in micromachinery and easy-to-use micro controller platforms, the uses of sensors have expanded beyond the most traditional fields of temperature, pressure or flow measurement, for example into MARG sensors. Moreover, analog sensors such as potentiometers and force-sensing resistors are still widely used. Applications include manufacturing and machinery, airplanes and aerospace, cars, medicine, and robotics.it is also included in our day-to-day life.

A sensor's sensitivity indicates how much the sensor's output changes when the input quantity being measured changes. For instance, if the mercury in a thermometer moves 1 cm when the temperature changes by 1 °C, the sensitivity is 1 cm/°C (it is basically the slope Dy/Dx assuming a linear characteristic). Some sensors can also affect what they measure; for instance, a room temperature thermometer inserted into a hot cup of liquid cools the liquid while the liquid heats the thermometer. Sensors need to be designed to have a small effect on what is measured; making the sensor smaller often improves this and may introduce other advantages. Technological progress allows more and more sensors to be manufactured on a microscopic scale as microsensors using MEMS technology. In most cases, a microsensor reaches a significantly higher speed and sensitivity compared with macroscopic approaches.

Classification of Measurement Errors

The sensitivity is then defined as the ratio between the output signal and measured property. For example, if a sensor measures temperature and has a voltage output, the sensitivity is a constant with the unit [V/K]; this sensor is linear because the ratio is constant at all points of measurement.

For an analog sensor signal to be processed, or used in digital equipment, it needs to be converted to a digital signal, using an analog-to-digital converter.

Sensor Deviations

If the sensor is not ideal, several types of deviations can be observed:

- The sensitivity may in practice differ from the value specified. This is called a sensitivity error.

- Since the range of the output signal is always limited, the output signal will eventually reach a minimum or maximum when the measured property exceeds the limits. The full scale range defines the maximum and minimum values of the measured property.

- If the output signal is not zero when the measured property is zero, the sensor has an offset or bias. This is defined as the output of the sensor at zero input.

- If the sensitivity is not constant over the range of the sensor, this is called nonlinearity. Usually, this is defined by the amount the output differs from ideal behavior over the full range of the sensor, often noted as a percentage of the full range.

- If the deviation is caused by a rapid change of the measured property over time, there is a dynamic error. Often, this behavior is described with a bode plot showing sensitivity error and phase shift as a function of the frequency of a periodic input signal.

- If the output signal slowly changes independent of the measured property, this is defined as drift (telecommunication). Long term drift usually indicates a slow degradation of sensor properties over a long period of time.

- Noise is a random deviation of the signal that varies in time.

- Hysteresis is an error caused by when the measured property reverses direction, but there is some finite lag in time for the sensor to respond, creating a different offset error in one direction than in the other.

- If the sensor has a digital output, the output is essentially an approximation of the measured property. The approximation error is also called digitization error.

- If the signal is monitored digitally, limitation of the sampling frequency also can cause a dynamic error, or if the variable or added noise changes periodically at a frequency near a multiple of the sampling rate may induce aliasing errors.

- The sensor may to some extent be sensitive to properties other than the property being measured. For example, most sensors are influenced by the temperature of their environment.

All these deviations can be classified as systematic errors or random errors. Systematic errors can sometimes be compensated for by means of some kind of calibration strategy. Noise is a random error that can be reduced by signal processing, such as filtering, usually at the expense of the dynamic behavior of the sensor.

Resolution

The resolution of a sensor is the smallest change it can detect in the quantity that it is measuring. Often in a digital display, the least significant digit will fluctuate, indicating that changes of that magnitude are only just resolved. The resolution is related to the precision with which the measurement is made. For example, a scanning tunneling probe (a fine tip near a surface collects an electron tunneling current) can resolve atoms and molecules.

Types

- Pressure sensor
- Ultrasonic sensor
- Humidity sensor
- Gas sensor
- PIR motion sensor
- Acceleration sensor
- Displacement sensor
- Force measurement sensor
- color sensor
- gyro sensor

Sensors in Nature

All living organisms contain biological sensors with functions similar to those of the mechanical devices described. Most of these are specialized cells that are sensitive to:

- Light, motion, temperature, magnetic fields, gravity, humidity, moisture, vibration, pressure, electrical fields, sound, and other physical aspects of the external environment
- Physical aspects of the internal environment, such as stretch, motion of the organism, and position of appendages (proprioception)
- Environmental molecules, including toxins, nutrients, and pheromones
- Estimation of biomolecules interaction and some kinetics parameters
- Internal metabolic indicators, such as glucose level, oxygen level, or osmolality
- Internal signal molecules, such as hormones, neurotransmitters, and cytokines

- • Differences between proteins of the organism itself and of the environment or alien creatures.

Chemical Sensor

A chemical sensor is a self-contained analytical device that can provide information about the chemical composition of its environment, that is, a liquid or a gas phase. The information is provided in the form of a measurable physical signal that is correlated with the concentration of a certain chemical species (termed as analyte). Two main steps are involved in the functioning of a chemical sensor, namely, recognition and transduction. In the recognition step, analyte molecules interact selectively with receptor molecules or sites included in the structure of the recognition element of the sensor. Consequently, a characteristic physical parameter varies and this variation is reported by means of an integrated transducer that generates the output signal. A chemical sensor based on recognition material of biological nature is a biosensor. However, as synthetic biomimetic materials are going to substitute to some extent recognition biomaterials, a sharp distinction between a biosensor and a standard chemical sensor is superfluous. Typical biomimetic materials used in sensor development are molecularly imprinted polymers and aptamers.

Biosensor

In biomedicine and biotechnology, sensors which detect analytes thanks to a biological component, such as cells, protein, nucleic acid or biomimetic polymers, are called biosensors. Whereas a non-biological sensor, even organic (=carbon chemistry), for biological analytes is referred to as sensor or nanosensor. This terminology applies for both in-vitro and in vivo applications. The encapsulation of the biological component in biosensors, presents a slightly different problem that ordinary sensors; this can either be done by means of a semipermeable barrier, such as a dialysis membrane or a hydrogel, or a 3D polymer matrix, which either physically constrains the sensing macromolecule or chemically constrains the macromolecule by bounding it to the scaffold.

Actuator

An actuator is a component of machines that is responsible for moving or controlling a mechanism or system.

An actuator requires a control signal and source of energy. The control signal is relatively low energy and may be electric voltage or current, pneumatic or hydraulic pressure, or even human power. The supplied main energy source may be electric current, hydraulic fluid pressure, or pneumatic pressure. When the control signal is received, the actuator responds by converting the energy into mechanical motion.

An actuator is the mechanism by which a control system acts upon an environment. The control system can be simple (a fixed mechanical or electronic system), software-based (e.g. a printer driver, robot control system), a human, or any other input.

History

The history of the pneumatic actuation system and the hydraulic actuation system dates to around the time of World War II (1938). It was first created by Xhiter Anckeleman (pronounced 'Ziter') who used his knowledge of engines and brake systems to come up with a new solution to ensure that the brakes on a car exert the maximum force, with the least possible wear and tear.

Hydraulic

A hydraulic actuator consists of cylinder or fluid motor that uses hydraulic power to facilitate mechanical operation. The mechanical motion gives an output in terms of linear, rotary or oscillatory motion. Because liquids are nearly impossible to compress, a hydraulic actuator can exert a large force. The drawback of this approach is its limited acceleration.

The hydraulic cylinder consists of a hollow cylindrical tube along which a piston can slide. The term *single acting* is used when the fluid pressure is applied to just one side of the piston. The piston can move in only one direction, a spring being frequently used to give the piston a return stroke. The term *double acting* is used when pressure is applied on each side of the piston; any difference in pressure between the two side of the piston moves the piston to one side or the other.

Pneumatic

A pneumatic actuator converts energy formed by vacuum or compressed air at high pressure into either linear or rotary motion. Pneumatic energy is desirable for main engine controls because it can quickly respond in starting and stopping as the power source does not need to be stored in reserve for operation.

Pneumatic rack and pinion actuators for valve controls of water pipes

Pneumatic actuators enable considerable forces to be produced from relatively small pressure changes. These forces are often used with valves to move diaphragms to affect the flow of liquid through the valve.

Electric

An electric actuator is powered by a motor that converts electrical energy into mechanical torque. The electrical energy is used to actuate equipment such as multi-turn valves. It is one of the cleanest and most readily available forms of actuator because it does not involve oil.

Thermal or Magnetic (Shape Memory Alloys)

Actuators which can be actuated by applying thermal or magnetic energy have been used in commercial applications. They tend to be compact, lightweight, economical and with high power density. These actuators use shape memory materials (SMMs), such as shape memory alloys (SMAs) or magnetic shape-memory alloys (MSMAs). Some popular manufacturers of these devices are Finnish Modti Inc., American Dynalloy and Rotork.

Mechanical

A mechanical actuator functions to execute movement by converting one kind of motion, such as rotary motion, into another kind, such as linear motion. An example is a rack and pinion. The operation of mechanical actuators is based on combinations of structural components, such as gears and rails, or pulleys and chains.

Examples and Applications

In engineering, actuators are frequently used as mechanisms to introduce motion, or to clamp an object so as to prevent motion. In electronic engineering, actuators are a subdivision of transducers. They are devices which transform an input signal (mainly an electrical signal) into some form of motion.

Examples of Actuators

- Comb drive
- Digital micromirror device
- Electric motor
- Electroactive polymer
- Hydraulic cylinder
- Piezoelectric actuator
- Pneumatic actuator
- Servomechanism
- Thermal bimorph
- Screw jack

Circular to Linear Conversion

Motors are mostly used when circular motions are needed, but can also be used for linear applications by transforming circular to linear motion with a lead screw or similar mechanism. On the other hand, some actuators are intrinsically linear, such as piezoelectric actuators. Conversion between circular and linear motion is commonly made via a few simple types of mechanism including:

- Screw: Screw jack, ball screw and roller screw actuators all operate on the principle of the simple machine known as the screw. By rotating the actuator's nut, the screw shaft moves in a line. By moving the screw shaft, the nut rotates.

- Wheel and axle: Hoist, winch, rack and pinion, chain drive, belt drive, rigid chain and rigid belt actuators operate on the principle of the wheel and axle. By rotating a wheel/axle (e.g. drum, gear, pulley or shaft) a linear member (e.g. cable, rack, chain or belt) moves. By moving the linear member, the wheel/axle rotates.

Virtual Instrumentation

In virtual instrumentation, actuators and sensors are the hardware complements of virtual instruments.

Performance Metrics

Performance metrics for actuators include speed, acceleration, and force (alternatively, angular speed, angular acceleration, and torque), as well as energy efficiency and considerations such as mass, volume, operating conditions, and durability, among others.

Force

When considering force in actuators for applications, two main metrics should be considered. These two are static and dynamic loads. Static load is the force capability of the actuator while not in motion. Conversely, the dynamic load of the actuator is the force capability while in motion. The two aspects rarely have the same weight capability and must be considered separately.

Speed

Speed should be considered primarily at a no-load pace, since the speed will invariably decrease as the load amount increases. The rate the speed will decrease will directly correlate with the amount of force and the initial speed.

Operating Conditions

Actuators are commonly rated using the standard IP Code rating system. Those that are rated for dangerous environments will have a higher IP rating than those for personal or common industrial use.

Durability

This will be determined by each individual manufacturer, depending on usage and quality.

References

- Bennett, S. (1993). A History of Control Engineering 1930–1955. London: Peter Peregrinus Ltd. on behalf of the Institution of Electrical Engineers. ISBN 0-86341-280-7.

- Bănică Florinel-Gabriel (2012). Chemical Sensors and Biosensors:Fundamentals and Applications. Chichester, UK: John Wiley & Sons. p. 576. ISBN 978-1-118-35423-0.

- "What's the Difference Between Pneumatic, Hydraulic, and Electrical Actuators?". machinedesign.com. Retrieved 2016-04-26.

Technologies Related to Sensors

Sensors are used in everyday objects. Some of the technologies related to sensors are explained in this chapter. Data logger, metal detectors, global positioning system, sonar, echo sounding and load cell are all technologies related to sensors. The topics discussed in the chapter are of great importance to broaden the existing knowledge on sensors.

Data Logger

A data logger (also datalogger or data recorder) is an electronic device that records data over time or in relation to location either with a built in instrument or sensor or via external instruments and sensors. Increasingly, but not entirely, they are based on a digital processor (or computer). They generally are small, battery powered, portable, and equipped with a microprocessor, internal memory for data storage, and sensors. Some data loggers interface with a personal computer, and use software to activate the data logger and view and analyze the collected data, while others have a local interface device (keypad, LCD) and can be used as a stand-alone device.

Data logger Cube storing technical and sensor data

Data loggers vary between general purpose types for a range of measurement applications to very specific devices for measuring in one environment or application type only. It is common for general purpose types to be programmable; however, many remain as static machines with only a limited number or no changeable parameters. Electronic data loggers have replaced chart recorders in many applications.

One of the primary benefits of using data loggers is the ability to automatically collect data on a 24-hour basis. Upon activation, data loggers are typically deployed and left unattended to measure and record information for the duration of the monitoring period. This allows for a comprehensive, accurate picture of the environmental conditions being monitored, such as air temperature and relative humidity.

The cost of data loggers has been declining over the years as technology improves and costs are reduced. Simple single channel data loggers cost as little as $25. More complicated loggers may costs hundreds or thousands of dollars.

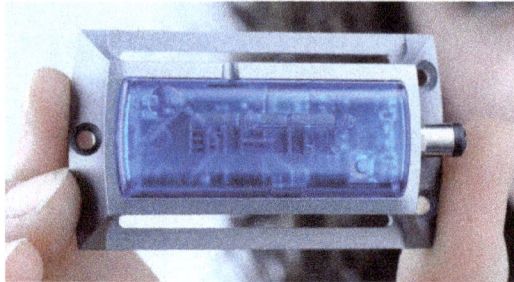

Small data logger with integrated sensors measuring temperature, pressure, humidity, light and 3-axis acceleration

Data Formats

Standardisation of protocols and data formats has been a problem but is now growing in the industry and XML, JSON, and YAML are increasingly being adopted for data exchange. The development of the Semantic Web and the Internet of Things is likely to accelerate this present trend.

Instrumentation Protocols

Several protocols have been standardised including a smart protocol, SDI-12, that allows some instrumentation to be connected to a variety of data loggers. The use of this standard has not gained much acceptance outside the environmental industry. SDI-12 also supports multi drop instruments. Some datalogging companies are also now supporting the MODBUS standard. This has been used traditionally in the industrial control area, and there are many industrial instruments which support this communication standard. Another multi drop protocol which is now starting to become more widely used is based upon Canbus (ISO 11898). Some data loggers use a flexible scripting environment to adapt themselves to various non-standard protocols.

Data Logging Versus Data Acquisition

The terms data logging and data acquisition are often used interchangeably. However, in a historical context they are quite different. A data logger is a data acquisition system, but a data acquisition system is not necessarily a data logger.

- Data loggers typically have slower sample rates. A maximum sample rate of 1 Hz may be considered to be very fast for a data logger, yet very slow for a typical data acquisition system.

- Data loggers are implicitly stand-alone devices, while typical data acquisition system must remain tethered to a computer to acquire data. This stand-alone aspect of data loggers implies on-board memory that is used to store acquired data. Sometimes this memory is very large to accommodate many days, or even months, of unattended recording. This memory may be battery-backed static random access memory, flash memory or EEPROM. Earlier data loggers used magnetic tape, punched paper tape, or directly viewable records such as "strip chart recorders".

- Given the extended recording times of data loggers, they typically feature a mechanism to record the date and time in a timestamp to ensure that each recorded data value is associated with a date and time of acquisition in order to produce a sequence of events. As such, data loggers typically employ built-in real-time clocks whose published drift can be an important consideration when choosing between data loggers.

- Data loggers range from simple single-channel input to complex multi-channel instruments. Typically, the simpler the device the less programming flexibility. Some more sophisticated instruments allow for cross-channel computations and alarms based on predetermined conditions. The newest of data loggers can serve web pages, allowing numerous people to monitor a system remotely.

- The unattended and remote nature of many data logger applications implies the need in some applications to operate from a DC power source, such as a battery. Solar power may be used to supplement these power sources. These constraints have generally led to ensure that the devices they market are extremely power efficient relative to computers. In many cases they are required to operate in harsh environmental conditions where computers will not function reliably.

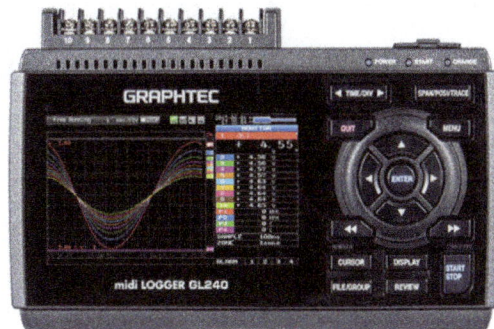

Portable Dataloggers may reach up to 20 channels with maximum 10ms (100Hz) sampling rate.

This unattended nature also dictates that data loggers must be extremely reliable. Since they may operate for long periods nonstop with little or no human supervision, and may be installed in harsh or remote locations, it is imperative that so long as they have power, they will not fail to log data for any reason. Manufacturers go to great length to ensure that the devices can be depended on in these applications. As such dataloggers are almost completely immune to the problems that might affect a general-purpose computer in the same application, such as program crashes and the instability of some operating systems.

Applications

Applications of data logging include:

- Unattended weather station recording (such as wind speed / direction, temperature, relative humidity, solar radiation).

- Unattended hydrographic recording (such as water level, water depth, water flow, water pH, water conductivity).

Data logger application for weather station at P2I LIPI

- Unattended soil moisture level recording.

- Unattended gas pressure recording.

- Offshore buoys for recording a variety of environmental conditions.

- Road traffic counting.

- Measure temperatures (humidity, etc.) of perishables during shipments: Cold chain.

- Measure variations in light intensity.

- Process monitoring for maintenance and troubleshooting applications.

- Process monitoring to verify warranty conditions.

- Wildlife research with pop-up archival tags.

- Measure vibration and handling shock (drop height) environment of distribution packaging.

- Tank level monitoring.

- Deformation monitoring of any object with geodetic or geotechnical sensors controlled by an automatic deformation monitoring system.

- Environmental monitoring.

- Vehicle Testing (including crash testing).

- Motor Racing.

- Monitoring of relay status in railway signalling.

- For science education enabling 'measurement', 'scientific investigation' and an appreciation of 'change'.

- Record trend data at regular intervals in veterinary vital signs monitoring.

- Load profile recording for energy consumption management.

- Temperature, humidity and power use for heating and air conditioning efficiency studies.

- Water level monitoring for groundwater studies.

- Digital electronic bus sniffer for debug and validation.

Examples

- Black-box (stimulus/response) loggers:

  o A flight data recorder (FDR), a piece of recording equipment used to collect specific aircraft performance data. The term may also be used, albeit less accurately, to describe the cockpit voice recorder (CVR), another type of data recording device found on board aircraft.

  o An event data recorder (EDR), a device installed by the manufacturer in some automobiles which collects and stores various data during the time-frame immediately before and after a crash.

  o A voyage data recorder (VDR), a data recording system designed to collect data from various sensors on board a ship.

  o A train event recorder, a device that records data about the operation of train controls and performance in response to those controls and other train control systems.

  o In automobiles, all diagnostic trouble codes (DTCs) are logged in engine control units (ECUs) so that at the time of service of a vehicle, a service engineer will read all the DTCs using Tech-2 or similar tools connected to the on-board diagnostics port, and will come to know problems occurred in the vehicle. Sometimes a small OBD data logger is plugged into the same port to continuously record vehicle data.

  o In embedded system and digital electronics design, specialized high-speed digital data logger help overcome the limitations of more traditional instruments such as the oscilloscope and the logic analyzer. The main advantage of a data logger is its ability to record very long traces, which proves very useful when trying to correct functional bugs that happen once in while.

  o In the racing industry, Data Loggers are used to record data such as braking points, lap/sector timing, and track maps, as well as any on-board vehicle sensors.

- Health data loggers:

  o The growing, preparation, storage and transportation of food. Data logger is generally used for data storage and these are small in size.

- o A Holter monitor is a portable device for continuously monitoring various electrical activity of the cardiovascular system for at least 24 hours.

- o Electronic health record loggers.

- Other general data acquisition loggers:

- o An (scientific) experimental testing data acquisition tool.

- o Ultra Wideband Data Recorder, high-speed data recording up to 2 GigaSamples per second.

- o An open source data logger based on the Raspberry Pi computer

Future Directions

Data Loggers are changing more rapidly now than ever before. The original model of a stand-alone data logger is changing to one of a device that collects data but also has access to wireless communications for alarming of events, automatic reporting of data and remote control. Data loggers are beginning to serve web pages for current readings, e-mail their alarms and FTP their daily results into databases or direct to the users. Very recently, there is a trend to move away from proprietary products with commercial software to open source software and hardware devices. The Raspberry Pi single-board computer is among others a popular platform hosting real-time Linux or preemptive-kernel Linux operating systems with many

- digital interfaces like I2C, SPI or UART enabling the direct interconnection of a digital sensor and a computer,

- and unlimited number of configurations to show measurements in real-time over the internet, process data, plot charts and diagrams...

There are more and more community-developed (open source) projects for data acquisition / data logging.

Metal Detector

A metal detector is an electronic instrument which detects the presence of metal nearby. Metal detectors are useful for finding metal inclusions hidden within objects, or metal objects buried underground. They often consist of a handheld unit with a sensor probe which can be swept over the ground or other objects. If the sensor comes near a piece of metal this is indicated by a changing tone in earphones, or a needle moving on an indicator. Usually the device gives some indication of distance; the closer the metal is, the higher the tone in the earphone or the higher the needle goes. Another common type are stationary "walk through" metal detectors used for security screening at access points in prisons, courthouses, and airports to detect concealed metal weapons on a person's body.

A U.S. Army soldier uses a metal detector to search for weapons and ammunition in March 2004.

The simplest form of a metal detector consists of an oscillator producing an alternating current that passes through a coil producing an alternating magnetic field. If a piece of electrically conductive metal is close to the coil, eddy currents will be induced in the metal, and this produces a magnetic field of its own. If another coil is used to measure the magnetic field (acting as a magnetometer), the change in the magnetic field due to the metallic object can be detected.

U.S. Army soldiers use a metal detector in 2002.

The first industrial metal detectors were developed in the 1960s and were used extensively for mineral prospecting and other industrial applications. Uses include de-mining (the detection of land mines), the detection of weapons such as knives and guns (especially in airport security), geophysical prospecting, archaeology and treasure hunting. Metal detectors are also used to detect foreign bodies in food, and in the construction industry to detect steel reinforcing bars in concrete and pipes and wires buried in walls and floors.

History and Development

Towards the end of the 19th century, many scientists and engineers used their growing knowledge of electrical theory in an attempt to devise a machine which would pinpoint metal. The use of such a device to find ore-bearing rocks would give a huge advantage to any miner who employed it. Early machines were crude, used a lot of battery power, and worked only to a very limited degree. In 1874, Parisian inventor Gustave Trouvé developed a hand-held device for locating and extracting metal objects such as bullets from human patients. Inspired by Trouvé, Alexander Graham Bell developed a similar device to attempt to locate a bullet lodged in the chest of American President James Garfield in 1881; the metal detector worked correctly but the attempt was unsuccessful because the metal coil spring bed Garfield was lying on confused the detector.

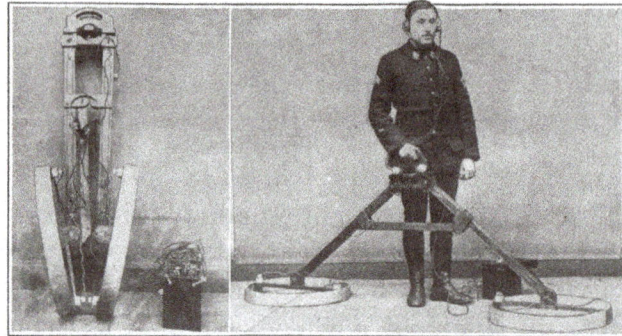

Early metal detector, 1919, used to find unexploded bombs in France after World War 1.

Modern Developments

The modern development of the metal detector began in the 1920s. Gerhard Fischer had developed a system of radio direction-finding, which was to be used for accurate navigation. The system worked extremely well, but Fischer noticed that there were anomalies in areas where the terrain contained ore-bearing rocks. He reasoned that if a radio beam could be distorted by metal, then it should be possible to design a machine which would detect metal using a search coil resonating at a radio frequency. In 1925 he applied for, and was granted, the first patent for a metal detector. Although Gerhard Fischer was the first person granted a patent for a metal detector, the first to apply was Shirl Herr, a businessman from Crawfordsville, Indiana. His application for a hand-held Hidden-Metal Detector was filed in February 1924, but not patented until July 1928. Herr assisted Italian leader Benito Mussolini in recovering items remaining from the Emperor Caligula's galleys at the bottom of Lake Nemi, Italy, in August 1929. Herr's invention was used by Admiral Richard Byrd's Second Antarctic Expedition in 1933, when it was used to locate objects left behind by earlier explorers. It was effective up to a depth of eight feet. However, it was one Lieutenant Józef Stanisław Kosacki, a Polish officer attached to a unit stationed in St Andrews, Fife, Scotland, during the early years of World War II, who refined the design into a practical Polish mine detector. These units were still quite heavy, as they ran on vacuum tubes, and needed separate battery packs.

The design invented by Kosacki was used extensively during the Second Battle of El Alamein when 500 units were shipped to Field Marshal Montgomery to clear the minefields of the retreating Germans, and later used during the Allied invasion of Sicily, the Allied invasion of Italy and the Invasion of Normandy.

As the creation and refinement of the device was a wartime military research operation, the knowledge that Kosacki created the first practical metal detector was kept secret for over 50 years.

Further Refinements

Many manufacturers of these new devices brought their own ideas to the market. White's Electronics of Oregon began in the 1950s by building a machine called the Oremaster Geiger Counter. Another leader in detector technology was Charles Garrett, who pioneered the BFO (Beat Frequency Oscillator) machine. With the invention and development of the transistor in the 1950s and 1960s, metal detector manufacturers and designers made smaller lighter machines with improved circuitry, running on small battery packs. Companies sprang up all over the USA and Britain to supply the growing demand.

Modern top models are fully computerized, using integrated circuit technology to allow the user to set sensitivity, discrimination, track speed, threshold volume, notch filters, etc., and hold these parameters in memory for future use. Compared to just a decade ago, detectors are lighter, deeper-seeking, use less battery power, and discriminate better.

Larger portable metal detectors are used by archaeologists and treasure hunters to locate metallic items, such as jewelry, coins, bullets, and other various artifacts buried shallowly underground.

Discriminators

The biggest technical change in detectors was the development of the induction-balance system. This system involved two coils that were electrically balanced. When metal was introduced to their vicinity, they would become unbalanced. What allowed detectors to discriminate between metals was the fact that every metal has a different phase response when exposed to alternating current. Scientists had long known of this fact by the time detectors were developed that could selectively detect desirable metals, while ignoring undesirable ones.

Even with discriminators, it was still a challenge to avoid undesirable metals, because some of them have similar phase responses e.g. tinfoil and gold, particularly in alloy form. Thus, improperly tuning out certain metals increased the risk of passing over a valuable find. Another disadvantage of discriminators was that they reduced the sensitivity of the machines.

New Coil Designs

Coil designers also tried out innovative designs. The original induction balance coil system consisted of two identical coils placed on top of one another. Compass Electronics produced a new design: two coils in a D shape, mounted back-to-back to form a circle. This system was widely used in the 1970s, and both concentric and D type (or widescan as they became known) had their fans. Another development was the invention of detectors which could cancel out the effect of mineralization in the ground. This gave greater depth, but was a non-discriminate mode. It worked best at lower frequencies than those used before, and frequencies of 3 to 20 kHz were found to produce the best results. Many detectors in the 1970s had a switch which enabled the user to switch between the discriminate mode and the non-discriminate mode. Later developments switched electronically between both modes. The development of the induction balance detector would ultimately result in the motion detector, which constantly checked and balanced the background mineralization.

Pulse Induction

At the same time, developers were looking at using a different technique in metal detection called pulse induction. Unlike the beat frequency oscillator or the induction balance machines which both used a uniform alternating current at a low frequency, the pulse induction machine simply magnetized the ground with a relatively powerful, momentary current through a search coil. In the absence of metal, the field decayed at a uniform rate, and the time it took to fall to zero volts could be accurately measured. However, if metal was present when the machine fired, a small eddy current would be induced in the metal, and the time for sensed current decay would be increased. These time differences were minute, but the improvement in electronics made it possible to measure them accurately and identify the presence of metal at a reasonable distance. These new machines had one major advantage: they were mostly impervious to the effects of mineralization, and rings and other jewelry could now be located even under highly mineralized black sand. The addition of computer control and digital signal processing have further improved pulse induction sensors.

A pulse induction metal detector with an array of coils

Uses

Archaeology

Metal detectors are widely used in archaeology with the first recorded use by military historian Don Rickey in 1958 who used one to detect the firing lines at Little Big Horn. However archaeologists oppose the use of metal detectors by "artifact seekers" or "site looters" whose activities disrupt archaeological sites.

England and Wales

In England and Wales metal detecting is legal provided that permission is granted by the landowner, and that the area is not a Scheduled Ancient Monument, a site of special scientific interest (SSSI), or covered by elements of the Countryside Stewardship Scheme.

Items discovered which fall within the definition of treasure must be reported to the coroner or a place designated by the coroner for treasure. The voluntary reporting of finds which do not qualify as treasure to the Portable Antiquities Scheme or the UK Detector Finds Database is encouraged.

Scotland

The situation in Scotland is very different. Under the Scots law principle of *bona vacantia*, the Crown has claim over any object of any material value where the original owner cannot be traced. There is also no 300 year limit to Scottish finds. Any artifact found, whether by metal detector survey or from an archaeological excavation, must be reported to the Crown through the Treasure Trove Advisory Panel at the National Museums of Scotland. The panel then determines what will happen to the artifacts. Reporting is not voluntary, and failure to report the discovery of historic artifacts is a criminal offence in Scotland.

France

The sale of metal detectors is allowed in France. The first use of metal detectors in France which led to archaeological discoveries occurred in 1958: people living in the city of Graincourt-lès-Havrincourt who were seeking copper from world war I bombshell with military mine detector found a Roman silver treasure. The French law on metal detecting is ambiguous because it refers only to the objective pursued by the user of a metal detector. The first law to regulate the use of metal detectors was Law No. 89-900 of 18 December 1989. This last is resumed without any change in Article L. 542-1 of the code of the heritage, which states that "no person may use the equipment for the detection of metal objects, for the purpose of research monuments and items of interest prehistory, history, art and archeology without having previously obtained a administrative authorization issued based on the applicant's qualification and the nature and method of research. " Outside the research of archaeological objects, using a metal detector does not require specific authorization, except that of the owner of the land. We often read, from some archaeologists, that the use of a metal detector is itself prohibited without official authorization. This is false. To realize this, one must look to the legislative intent in enacting the Law No. 89-900 of 18 December 1989. Asked about Law No. 89-900 of 18 December 1989 by the member of parliament mister Calloud, Jack Lang, Minister of Culture at the time, replied by letter the following: "The new law does not prohibit the use of metal detectors but only regulates the use. If the purpose of such use is the search for archaeological remains, prior authorization is required from my services. Apart from this case, the law ask to be reported to the appropriate authorities an accidental discovery of archaeological remains." The entire letter of Jack Lang was published in 1990 in a French metal detection magazine, and then, to be visible on internet, scanned with permission of the author of the magazine on a French metal detection website.

As a Hobby

This 156-troy-ounce (4.9 kg) gold nugget, known as the Mojave Nugget, was found by an individual prospector in the Southern California Desert using a metal detector

There are various types of hobby activities involving metal detectors:

- Coin shooting is looking for coins after an event involving many people, like a baseball game, or simply looking for any old coins. Some coin shooters conduct historical research to locate sites with potential to give up historical and collectible coins.

- Prospecting is looking for valuable metals like gold, silver, and copper in their natural forms, such as nuggets or flakes.

- General metal detecting is very similar to coin shooting except that the user is after any type of historical artifact. Detector users may be dedicated to preserving historical artifacts, and often have considerable expertise. Coins, bullets, buttons, axe heads, and buckles are just a few of the items that are commonly found by relic hunters; in general the potential is far greater in Europe and Asia than in many other parts of the world. More valuable finds in Britain alone include the Staffordshire Hoard of Anglo-Saxon gold, sold for £3,285,000, the gold Celtic Newark Torc, the Ringlemere Cup, West Bagborough Hoard, Milton Keynes Hoard, Roman Crosby Garrett Helmet, Stirling Hoard, Collette Hoard and thousands of smaller finds.

- Beach combing is hunting for lost coins or jewelry on a beach. Beach hunting can be as simple or as complicated as one wishes to make it. Many dedicated beach hunters also familiarize themselves with tide movements and beach erosion.

- Metal detecting clubs across the United States, United Kingdom and Canada exist for hobbyists to learn from others, show off finds from their hunts and to learn more about the hobby.

Politics and Conflicts in The Metal Detecting Hobby in The USA

The metal detecting community and professional archaeologists have different ideas related to the recovery and preservation of historic finds and locations. Archaeologists claim that detector hobbyists take an artifact-centric approach, removing these from their context resulting in a permanent loss of historical information. Archaeological looting of places like Slack Farm in 1987 and Petersburg National Battlefield serve as evidence against allowing unsupervised metal detecting in historic locations.

Hobbyists often state that professional archaeologists' resource limitations results in the loss or damage of many artifacts by plows, development, erosion and livestock. The language and breadth of legislation regarding artifact collection is also an issue, as the Archaeological Resources Protection Act of 1979 excludes scattered coins, the main target of inland hobbyists. Many hobbyists take issue with the breadth of metal detecting bans, marking large swaths of property off-limits which are either well-documented already or unlikely to ever receive professional attention. Suggestions to certify or offer limited permits for detecting at historic sites have been attempted in some areas of the United States.

Recently, productive efforts for cooperation between professionals and metal detecting hobbyists have begun, including the Montpelier Archeology Project and Battlefield Restoration and Archaeological Volunteer Organization (BRAVO) and many more. In these programs, skilled detector hobbyists work with experienced professionals with common goals of accurate, efficient site discovery and excavation. Away from supervised sites, hobbyists using improved record keeping and employment of global positioning system, GIS, logbooks, photo scales and online databases may aid professionals in evaluating possible sites. When searching for a site, hobbyists can aid with electronic scanning, reducing the need for test holes. Some land managers, such as the Tennessee Valley Authority have cited a role for amateur archaeologists in protecting sensitive sites from illegal looting and metal detector hobbyists have aided in the location and preservation of many sites.

Security Screening

A series of aircraft hijackings led the United States in 1972 to adopt metal detector technology to screen airline passengers, initially using magnetometers that were originally designed for logging operations to detect spikes in trees. The Finnish company Outokumpu adapted mining metal detectors in the 1970s, still housed in a large cylindrical pipe, to make a commercial walk-through security detector. The development of these systems continued in a spin-off company and systems branded as Metor Metal Detectors evolved in the form of the rectangular gantry now standard in airports. In common with the developments in other uses of metal detectors both alternating current and pulse systems are used, and the design of the coils and the electronics has moved forward to improve the discrimination of these systems. In 1995 systems such as the Metor 200 appeared with the ability to indicate the approximate height of the metal object above the ground, enabling security personnel to more rapidly locate the source of the signal. Smaller hand held metal detectors are also used to locate a metal object on a person more precisely.

Metal detectors at Berlin Schönefeld Airport

Industrial Metal Detectors

Industrial metal detectors are used in the pharmaceutical, food, beverage, textile, garment, plastics, chemicals, lumber, mining, and packaging industries.

Contamination of food by metal shards from broken processing machinery during the manufacturing process is a major safety issue in the food industry. Metal detectors for this purpose are widely used and integrated into the production line.

Current practice at garment or apparel industry plants is to apply metal detecting after the garments are completely sewn and before garments are packed to check whether there is any metal contamination (needle, broken needle, etc.) in the garments. This needs to be done for safety reasons.

The industrial metal detector was developed by Bruce Kerr and David Hiscock in 1947. The founding company Goring Kerr pioneered the use and development of the first industrial metal detector. Mars Incorporated was one of the first customers of Goring Kerr using their Metlokate metal detector to inspect Mars bars.

The basic principle of operation for the common industrial metal detector is based on a 3 coil design. This design utilizes an AM (amplitude modulated) transmitting coil and two receiving coils one on either side of the transmitter. The design and physical configuration of the receiving coils are instrumental in the ability to detect very small metal contaminates of 1 mm or smaller. Today modern metal detectors continue to utilize this configuration for the detection of tramp metal.

The coil configuration is such that it creates an opening whereby the product (food, plastics, pharmaceuticals, etc.) passes through the coils. This opening or aperture allows the product to enter and exit through the three coil system producing an equal but mirrored signal on the two receiving coils. The resulting signals are summed together effectively nullifying each other. Fortress Technology innovated a new feature, that allows the coil structure of their BSH Model to ignore the effects of vibration, even when inspecting conductive products.

When a metal contaminant is introduced into the product an unequal disturbance is created. This then creates a very small electronic signal. After suitable amplification a mechanical device mounted to the conveyor system is signaled to remove the contaminated product from the production line. This process is completely automated and allows manufacturing to operate uninterrupted.

Civil Engineering

In civil engineering, special metal detectors (cover meters) are used to locate reinforcement bars inside walls.

Photoelectric Sensor

A photoelectric sensor, or photo eye, is an equipment used to discover the distance, absence, or presence of an object by using a light transmitter, often infrared, and a photoelectric receiver. They are largely used in industrial manufacturing. There are three different useful types: opposed (through beam), retro-reflective, and proximity-sensing (diffused).

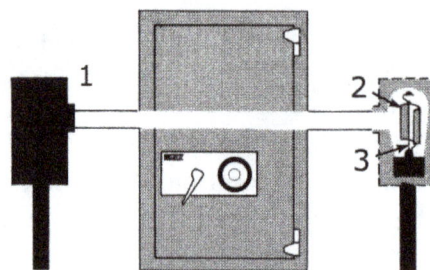

Conceptual through-beam system to detect unauthorized access to a secure door. If the beam is damaged, the detector triggers an alarm .

Types

A self-contained photoelectric sensor contains the optics, along with the electronics. It requires only a power source. The sensor performs its own modulation, demodulation, amplification, and output switching. Some self-contained sensors provide such options as built-in control timers or counters. Because of technological progress, self-contained photoelectric sensors have become increasingly smaller.

Remote photoelectric sensors used for remote sensing contain only the optical components of a sensor. The circuitry for power input, amplification, and output switching are located elsewhere, typically in a control panel. This allows the sensor, itself, to be very small. Also, the controls for the sensor are more accessible, since they may be bigger.

When space is restricted or the environment too hostile even for remote sensors, fiber optics may be used. Fiber optics are passive mechanical sensing components. They may be used with either remote or self-contained sensors. They have no electrical circuitry and no moving parts, and can safely pipe light into and out of hostile environments.

Sensing Modes

A through beam arrangement consists of a receiver located within the line-of-sight of the transmitter. In this mode, an object is detected when the light beam is blocked from getting to the receiver from the transmitter.

A retroreflective arrangement places the transmitter and receiver at the same location and uses a reflector to bounce the light beam back from the transmitter to the receiver. An object is sensed when the beam is interrupted and fails to reach the receiver.

A proximity-sensing (diffused) arrangement is one in which the transmitted radiation must reflect off the object in order to reach the receiver. In this mode, an object is detected when the receiver sees the transmitted source rather than when it fails to see it. As in retro-reflective sensors, diffuse sensor emitters and receivers are located in the same housing. But the target acts as the reflector, so that detection of light is reflected off the disturbance object. The emitter sends out a beam of light (most often a pulsed infrared, visible red, or laser) that diffuses in all directions, filling a detection area. The target then enters the area and deflects part of the beam back to the receiver. Detection occurs and output is turned on or off when sufficient light falls on the receiver.

Some photo eyes have two different operational types, light operate and dark operate. Light operate photo eyes become operational when the receiver "receives" the transmitter signal. Dark operate photo eyes become operational when the receiver "does not receive" the transmitter signal.

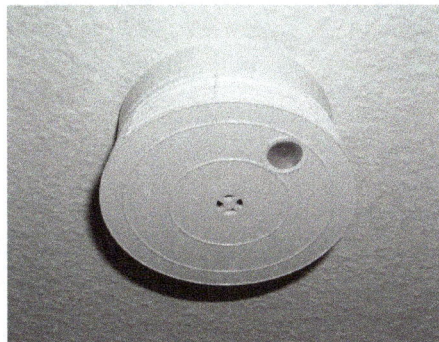

Certain types of smoke detector use a photoelectric sensor to warn of smoldering fires.

The detecting range of a photoelectric sensor is its "field of view", or the maximum distance from which the sensor can retrieve information, minus the minimum distance. A minimum detectable object is the smallest object the sensor can detect. More accurate sensors can often have minimum detectable objects of minuscule size.

Global Positioning System

The Global Positioning System (GPS), also known as Navstar, is a global navigation satellite system (GNSS) that provides location and time information in all weather conditions, anywhere on or near the Earth where there is an unobstructed line of sight to four or more GPS satellites. The GPS system operates independently of any telephonic or internet reception, though these technologies can enhance the usefulness of the GPS positioning information. The GPS system provides critical positioning capabilities to military, civil, and commercial users around the world. The United States government created the system, maintains it, and makes it freely accessible to anyone with a GPS receiver.

Artist's conception of GPS Block II-F satellite in Earth orbit.

The United States began the GPS project in 1973 to overcome the limitations of previous navigation systems, integrating ideas from several predecessors, including a number of classified engineering design studies from the 1960s. The U.S. Department of Defense (DoD) developed the system, which originally used 24 satellites. It became fully operational in 1995. Roger L. Easton, Ivan A. Getting and Bradford Parkinson of the Applied Physics Laboratory are credited with inventing it.

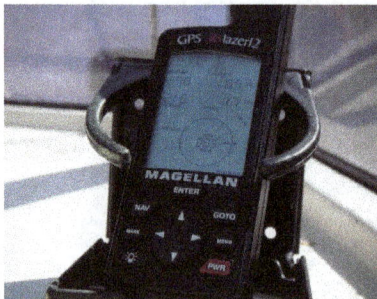

Civilian GPS receivers ("GPS navigation device") in a marine application.

Advances in technology and new demands on the existing system have now led to efforts to modernize the GPS and implement the next generation of GPS Block IIIA satellites and Next Generation Operational Control System (OCX). Announcements from Vice President Al Gore and the White House in 1998 initiated these changes. In 2000, the U.S. Congress authorized the modernization effort, GPS III.

In addition to GPS, other systems are in use or under development. The Russian Global Navigation Satellite System (GLONASS) was developed contemporaneously with GPS, but suffered from incomplete coverage of the globe until the mid-2000s. There are also the planned European Union Galileo positioning system, China's BeiDou Navigation Satellite System, the Japanese Quasi-Zenith Satellite System, and India's Indian Regional Navigation Satellite System.

History

The design of GPS is based partly on similar ground-based radio-navigation systems, such as LO-RAN and the Decca Navigator, developed in the early 1940s and used by the British Royal Navy during World War II.

In 1956, the German-American physicist Friedwardt Winterberg proposed a test of general relativity — detecting time slowing in a strong gravitational field using accurate atomic clocks placed in orbit inside artificial satellites.

Special and general relativity predict that the clocks on the GPS satellites would be seen by the Earth's observers to run 38 microseconds faster per day than the clocks on the Earth. The GPS calculated positions would quickly drift into error, accumulating to 10 kilometers per day. The relativistic time effect of the GPS clocks running faster than the clocks on earth was corrected for in the design of GPS.

Predecessors

The Soviet Union launched the first man-made satellite, Sputnik 1, in 1957. Two American physicists, William Guier and George Weiffenbach, at Johns Hopkins's Applied Physics Laboratory (APL), decided to monitor Sputnik's radio transmissions. Within hours they realized that, because of the Doppler effect, they could pinpoint where the satellite was along its orbit. The Director of the APL gave them access to their UNIVAC to do the heavy calculations required.

The next spring, Frank McClure, the deputy director of the APL, asked Guier and Weiffenbach to investigate the inverse problem — pinpointing the user's location, given that of the satellite. (At the time, the Navy was developing the submarine-launched Polaris missile, which required them to know the submarine's location.) This led them and APL to develop the TRANSIT system. In 1959, ARPA (renamed DARPA in 1972) also played a role in TRANSIT.

Official logo for NAVSTAR GPS

Emblem of the 50th Space Wing

The first satellite navigation system, TRANSIT, used by the United States Navy, was first successfully tested in 1960. It used a constellation of five satellites and could provide a navigational fix approximately once per hour.

In 1967, the U.S. Navy developed the Timation satellite that proved the ability to place accurate clocks in space, a technology required by GPS.

In the 1970s, the ground-based OMEGA navigation system, based on phase comparison of signal transmission from pairs of stations, became the first worldwide radio navigation system. Limitations of these systems drove the need for a more universal navigation solution with greater accuracy.

While there were wide needs for accurate navigation in military and civilian sectors, almost none of those was seen as justification for the billions of dollars it would cost in research, development, deployment, and operation for a constellation of navigation satellites. During the Cold War arms race, the nuclear threat to the existence of the United States was the one need that did justify this cost in the view of the United States Congress. This deterrent effect is why GPS was funded. It is also the reason for the ultra secrecy at that time. The nuclear triad consisted of the United States Navy's submarine-launched ballistic missiles (SLBMs) along with United States Air Force (USAF) strategic bombers and intercontinental ballistic missiles (ICBMs). Considered vital to the nuclear deterrence posture, accurate determination of the SLBM launch position was a force multiplier.

Precise navigation would enable United States ballistic missile submarines to get an accurate fix of their positions before they launched their SLBMs. The USAF, with two thirds of the nuclear triad, also had requirements for a more accurate and reliable navigation system. The Navy and Air Force were developing their own technologies in parallel to solve what was essentially the same problem.

To increase the survivability of ICBMs, there was a proposal to use mobile launch platforms (such as Russian SS-24 and SS-25) and so the need to fix the launch position had similarity to the SLBM situation.

In 1960, the Air Force proposed a radio-navigation system called MOSAIC (MObile System for Accurate ICBM Control) that was essentially a 3-D LORAN. A follow-on study, Project 57, was worked in 1963 and it was "in this study that the GPS concept was born." That same year, the concept was pursued as Project 621B, which had "many of the attributes that you now see in GPS" and promised increased accuracy for Air Force bombers as well as ICBMs.

Updates from the Navy TRANSIT system were too slow for the high speeds of Air Force operation. The Naval Research Laboratory continued advancements with their Timation (Time Navigation) satellites, first launched in 1967, and with the third one in 1974 carrying the first atomic clock into orbit.

Another important predecessor to GPS came from a different branch of the United States military. In 1964, the United States Army orbited its first Sequential Collation of Range (SECOR) satellite used for geodetic surveying. The SECOR system included three ground-based transmitters from known locations that would send signals to the satellite transponder in orbit. A fourth ground-based station, at an undetermined position, could then use those signals to fix its location precisely. The last SECOR satellite was launched in 1969.

Decades later, during the early years of GPS, civilian surveying became one of the first fields to make use of the new technology, because surveyors could reap benefits of signals from the less-than-complete GPS constellation years before it was declared operational. GPS can be thought of as an evolution of the SECOR system where the ground-based transmitters have been migrated into orbit.

Development

With these parallel developments in the 1960s, it was realized that a superior system could be developed by synthesizing the best technologies from 621B, Transit, Timation, and SECOR in a multi-service program.

During Labor Day weekend in 1973, a meeting of about twelve military officers at the Pentagon discussed the creation of a *Defense Navigation Satellite System (DNSS)*. It was at this meeting that the real synthesis that became GPS was created. Later that year, the DNSS program was named *Navstar*, or Navigation System Using Timing and Ranging. With the individual satellites being associated with the name Navstar (as with the predecessors Transit and Timation), a more fully encompassing name was used to identify the constellation of Navstar satellites, *Navstar-GPS*. Ten "Block I" prototype satellites were launched between 1978 and 1985 (with one prototype being destroyed in a launch failure).

After Korean Air Lines Flight 007, a Boeing 747 carrying 269 people, was shot down in 1983 after straying into the USSR's prohibited airspace, in the vicinity of Sakhalin and Moneron Islands, President Ronald Reagan issued a directive making GPS freely available for civilian use, once it was sufficiently developed, as a common good. The first Block II satellite was launched on February 14, 1989, and the 24th satellite was launched in 1994. The GPS program cost at this point, not including the cost of the user equipment, but including the costs of the satellite launches, has been estimated at about USD$5 billion (then-year dollars). Roger L. Easton is widely credited as the primary inventor of GPS.

Initially, the highest quality signal was reserved for military use, and the signal available for civilian use was intentionally degraded (Selective Availability). This changed with President Bill Clinton signing a policy directive in 1996 to turn off Selective Availability in May 2000 to provide the same precision to civilians that was afforded to the military. The directive was proposed by the U.S. Secretary of Defense, William Perry, because of the widespread growth of differential GPS services to improve civilian accuracy and eliminate the U.S. military advantage. Moreover, the U.S. military was actively developing technologies to deny GPS service to potential adversaries on a regional basis.

Since its deployment, the U.S. has implemented several improvements to the GPS service including new signals for civil use and increased accuracy and integrity for all users, all the while maintaining compatibility with existing GPS equipment. Modernization of the satellite system has been an ongoing initiative by the U.S. Department of Defense through a series of satellite acquisitions to meet the growing needs of the military, civilians, and the commercial market.

As of early 2015, high-quality, FAA grade, Standard Positioning Service (SPS) GPS receivers provide horizontal accuracy of better than 3.5 meters, although many factors such as receiver quality and atmospheric issues can affect this accuracy.

GPS is owned and operated by the United States Government as a national resource. The Department of Defense is the steward of GPS. *Interagency GPS Executive Board (IGEB)* oversaw GPS policy matters from 1996 to 2004. After that the National Space-Based Positioning, Navigation and Timing Executive Committee was established by presidential directive in 2004 to advise and coordinate federal departments and agencies on matters concerning the GPS and related systems. The executive committee is chaired jointly by the deputy secretaries of defense and transportation. Its membership includes equivalent-level officials from the departments of state, commerce, and homeland security, the joint chiefs of staff, and NASA. Components of the executive office of the president participate as observers to the executive committee, and the FCC chairman participates as a liaison.

The U.S. Department of Defense is required by law to "maintain a Standard Positioning Service (as defined in the federal radio navigation plan and the standard positioning service signal specification) that will be available on a continuous, worldwide basis," and "develop measures to prevent hostile use of GPS and its augmentations without unduly disrupting or degrading civilian uses."

Timeline and Modernization

Summary of satellites						
Block	Launch Period	Satellite launches				Currently in orbit and healthy
		Success	Failure	In preparation	Planned	
I	1978–1985	10	1	0	0	0
II	1989–1990	9	0	0	0	0
IIA	1990–1997	19	0	0	0	0
IIR	1997–2004	12	1	0	0	12
IIR-M	2005–2009	8	0	0	0	7
IIF	2010–2016	12	0	0	0	12
IIIA	From 2017	0	0	0	12	0
IIIB	—	0	0	0	8	0
IIIC	—	0	0	0	16	0
Total		70	2	0	36	31
(Last update: March 9, 2016) 8 satellites from Block IIA are placed in reserve USA-203 from Block IIR-M is unhealthy For a more complete list, see *list of GPS satellite launches*						

- In 1972, the USAF Central Inertial Guidance Test Facility (Holloman AFB), conducted developmental flight tests of two prototype GPS receivers over White Sands Missile Range, using ground-based pseudo-satellites.

- In 1978, the first experimental Block-I GPS satellite was launched.

- In 1983, after Soviet interceptor aircraft shot down the civilian airliner KAL 007 that strayed into prohibited airspace because of navigational errors, killing all 269 people on board, U.S. President Ronald Reagan announced that GPS would be made available for civilian uses once it was completed, although it had been previously published [in Navigation magazine] that the CA code (Coarse Acquisition code) would be available to civilian users.

- By 1985, ten more experimental Block-I satellites had been launched to validate the concept.

- Beginning in 1988, Command & Control of these satellites was transitioned from Onizuka AFS, California to the 2nd Satellite Control Squadron (2SCS) located at Falcon Air Force Station in Colorado Springs, Colorado.

- On February 14, 1989, the first modern Block-II satellite was launched.

- The Gulf War from 1990 to 1991 was the first conflict in which the military widely used GPS.

- In 1991, a project to create a miniature GPS receiver successfully ended, replacing the previous 23 kg military receivers with a 1.25 kg handheld receiver.

- In 1992, the 2nd Space Wing, which originally managed the system, was inactivated and replaced by the 50th Space Wing.

- By December 1993, GPS achieved initial operational capability (IOC), indicating a full constellation (24 satellites) was available and providing the Standard Positioning Service (SPS).

- Full Operational Capability (FOC) was declared by Air Force Space Command (AFSPC) in April 1995, signifying full availability of the military's secure Precise Positioning Service (PPS).

- In 1996, recognizing the importance of GPS to civilian users as well as military users, U.S. President Bill Clinton issued a policy directive declaring GPS a dual-use system and establishing an Interagency GPS Executive Board to manage it as a national asset.

- In 1998, United States Vice President Al Gore announced plans to upgrade GPS with two new civilian signals for enhanced user accuracy and reliability, particularly with respect to aviation safety and in 2000 the United States Congress authorized the effort, referring to it as *GPS III*.

- On May 2, 2000 "Selective Availability" was discontinued as a result of the 1996 executive order, allowing users to receive a non-degraded signal globally.

- In 2004, the United States Government signed an agreement with the European Community establishing cooperation related to GPS and Europe's planned Galileo system.

- In 2004, United States President George W. Bush updated the national policy and replaced the executive board with the National Executive Committee for Space-Based Positioning, Navigation, and Timing.

- November 2004, Qualcomm announced successful tests of assisted GPS for mobile phones.

- In 2005, the first modernized GPS satellite was launched and began transmitting a second civilian signal (L2C) for enhanced user performance.

- On September 14, 2007, the aging mainframe-based Ground Segment Control System was transferred to the new Architecture Evolution Plan.

- On May 19, 2009, the United States Government Accountability Office issued a report warning that some GPS satellites could fail as soon as 2010.

- On May 21, 2009, the Air Force Space Command allayed fears of GPS failure saying "There's only a small risk we will not continue to exceed our performance standard."

- On January 11, 2010, an update of ground control systems caused a software incompatibility with 8000 to 10000 military receivers manufactured by a division of Trimble Navigation Limited of Sunnyvale, Calif.

- On February 25, 2010, the U.S. Air Force awarded the contract to develop the GPS Next Generation Operational Control System (OCX) to improve accuracy and availability of GPS navigation signals, and serve as a critical part of GPS modernization.

Awards

On February 10, 1993, the National Aeronautic Association selected the GPS Team as winners of the 1992 Robert J. Collier Trophy, the nation's most prestigious aviation award. This team combines researchers from the Naval Research Laboratory, the USAF, the Aerospace Corporation, Rockwell International Corporation, and IBM Federal Systems Company. The citation honors them "for the most significant development for safe and efficient navigation and surveillance of air and spacecraft since the introduction of radio navigation 50 years ago."

Two GPS developers received the National Academy of Engineering Charles Stark Draper Prize for 2003:

- Ivan Getting, emeritus president of The Aerospace Corporation and an engineer at the Massachusetts Institute of Technology, established the basis for GPS, improving on the World War II land-based radio system called LORAN (*L*ong-range *R*adio *A*id to *N*avigation).

- Bradford Parkinson, professor of aeronautics and astronautics at Stanford University, conceived the present satellite-based system in the early 1960s and developed it in conjunction with the U.S. Air Force. Parkinson served twenty-one years in the Air Force, from 1957 to 1978, and retired with the rank of colonel.

GPS developer Roger L. Easton received the National Medal of Technology on February 13, 2006.

Francis X. Kane (Col. USAF, ret.) was inducted into the U.S. Air Force Space and Missile Pioneers Hall of Fame at Lackland A.F.B., San Antonio, Texas, March 2, 2010 for his role in space technology development and the engineering design concept of GPS conducted as part of Project 621B.

In 1998, GPS technology was inducted into the Space Foundation Space Technology Hall of Fame.

On October 4, 2011, the International Astronautical Federation (IAF) awarded the Global Positioning System (GPS) its 60th Anniversary Award, nominated by IAF member, the American Institute for Aeronautics and Astronautics (AIAA). The IAF Honors and Awards Committee recognized the uniqueness of the GPS program and the exemplary role it has played in building international collaboration for the benefit of humanity.

Basic Concept of GPS

Fundamentals

The GPS concept is based on time and the known position of specialized satellites. The satellites carry very stable atomic clocks that are synchronized to each other and to ground clocks. Any drift from true time maintained on the ground is corrected daily. Likewise, the satellite locations are known with great precision. GPS receivers have clocks as well; however, they are not synchronized with true time, and are less stable. GPS satellites continuously transmit their current time and position. A GPS receiver monitors multiple satellites and solves equations to determine the precise position of the receiver and its deviation from true time. At a minimum, four satellites must be in view of the receiver for it to compute four unknown quantities (three position coordinates and clock deviation from satellite time).

More Detailed Description

Each GPS satellite continually broadcasts a signal (carrier wave with modulation) that includes:

- A pseudorandom code (sequence of ones and zeros) that is known to the receiver. By time-aligning a receiver-generated version and the receiver-measured version of the code, the time of arrival (TOA) of a defined point in the code sequence, called an epoch, can be found in the receiver clock time scale

- A message that includes the time of transmission (TOT) of the code epoch (in GPS system time scale) and the satellite position at that time

Conceptually, the receiver measures the TOAs (according to its own clock) of four satellite signals. From the TOAs and the TOTs, the receiver forms four time of flight (TOF) values, which are (given the speed of light) approximately equivalent to receiver-satellite range differences. The receiver then computes its three-dimensional position and clock deviation from the four TOFs.

In practice the receiver position (in three dimensional Cartesian coordinates with origin at the Earth's center) and the offset of the receiver clock relative to the GPS time are computed simultaneously, using the navigation equations to process the TOFs.

The receiver's Earth-centered solution location is usually converted to latitude, longitude and height relative to an ellipsoidal Earth model. The height may then be further converted to height relative the geoid (e.g., EGM96) (essentially, mean sea level). These coordinates may be displayed, e.g. on a moving map display and/or recorded and/or used by some other system (e.g., a vehicle guidance system).

User-satellite Geometry

Although usually not formed explicitly in the receiver processing, the conceptual time differences of arrival (TDOAs) define the measurement geometry. Each TDOA corresponds to a hyperboloid of revolution. The line connecting the two satellites involved (and its exten-sions) forms the axis of the hyperboloid. The receiver is located at the point where three hyperbo-loids intersect.

It is sometimes incorrectly said that the user location is at the intersection of three spheres. While simpler to visualize, this is only the case if the receiver has a clock synchronized with the satellite clocks (i.e., the receiver measures true ranges to the satellites rather than range differences). There are significant performance benefits to the user carrying a clock synchronized with the satellites. Foremost is that only three satellites are needed to compute a position solution. If this were part of the GPS system concept so that all users needed to carry a synchronized clock, then a smaller number of satellites could be deployed. However, the cost and complexity of the user equipment would increase significantly.

Receiver in Continuous Operation

The description above is representative of a receiver start-up situation. Most receivers have a track algorithm, sometimes called a *tracker*, that combines sets of satellite measurements collected at different times—in effect, taking advantage of the fact that successive receiver positions are usually close to each other. After a set of measurements are processed, the tracker predicts the receiver location corresponding to the next set of satellite measurements. When the new measurements are collected, the receiver uses a weighting scheme to combine the new measurements with the tracker prediction. In general, a tracker can (a) improve receiver position and time accuracy, (b) reject bad measurements, and (c) estimate receiver speed and direction.

The disadvantage of a tracker is that changes in speed or direction can only be computed with a delay, and that derived direction becomes inaccurate when the distance traveled between two position measurements drops below or near the random error of position measurement. GPS units can use measurements of the Doppler shift of the signals received to compute velocity accurately. More advanced navigation systems use additional sensors like a compass or an inertial navigation system to complement GPS.

Non-Navigation Applications

In typical GPS operation as a navigator, four or more satellites must be visible to obtain an accurate result. The solution of the navigation equations gives the position of the receiver along with the difference between the time kept by the receiver's on-board clock and the true time-of-day, thereby eliminating the need for a more precise and possibly impractical receiver based clock. Applications for GPS such as time transfer, traffic signal timing, and synchronization of cell phone base stations, make use of this cheap and highly accurate timing. Some GPS applications use this time for display, or, other than for the basic position calculations, do not use it at all.

Although four satellites are required for normal operation, fewer apply in special cases. If one variable is already known, a receiver can determine its position using only three satellites. For example, a ship or aircraft may have known elevation. Some GPS receivers may use additional clues or assumptions such as reusing the last known altitude, dead reckoning, inertial navigation, or including information from the vehicle computer, to give a (possibly degraded) position when fewer than four satellites are visible.

Structure

The current GPS consists of three major segments. These are the space segment (SS), a control segment (CS), and a user segment (US). The U.S. Air Force develops, maintains, and operates the space and control segments. GPS satellites broadcast signals from space, and each GPS receiver uses these signals to calculate its three-dimensional location (latitude, longitude, and altitude) and the current time.

The space segment is composed of 24 to 32 satellites in medium Earth orbit and also includes the payload adapters to the boosters required to launch them into orbit. The control segment is composed of a master control station (MCS), an alternate master control station, and a host of dedicated and shared ground antennas and monitor stations. The user segment is composed of hundreds of thousands of U.S. and allied military users of the secure GPS Precise Positioning Service, and hundreds of millions of civil, commercial, and scientific users of the Standard Positioning Service .

Space Segment

The space segment (SS) is composed of the orbiting GPS satellites, or Space Vehicles (SV) in GPS parlance. The GPS design originally called for 24 SVs, eight each in three approximately circular orbits, but this was modified to six orbital planes with four satellites each. The six orbit planes have approximately 55° inclination (tilt relative to the Earth's equator) and are separated by 60° right ascension of the ascending node (angle along the equator from a reference point to the orbit's intersection). The orbital period is one-half a sidereal day, i.e., 11 hours and 58 minutes so that the satellites pass over the same locations or almost the same locations every day. The orbits are arranged so that at least six satellites are always within line of sight from almost everywhere on the Earth's surface. The result of this objective is that the four satellites are not evenly spaced (90 degrees) apart within each orbit. In general terms, the angular difference between satellites in each orbit is 30, 105, 120, and 105 degrees apart, which sum to 360 degrees.

Unlaunched GPS block II-A satellite on display at the San Diego Air & Space Museum

Orbiting at an altitude of approximately 20,200 km (12,600 mi); orbital radius of approximately 26,600 km (16,500 mi), each SV makes two complete orbits each sidereal day, repeating the same ground track each day. This was very helpful during development because even with only four satellites, correct alignment means all four are visible from one spot for a few hours each day. For military operations, the ground track repeat can be used to ensure good coverage in combat zones.

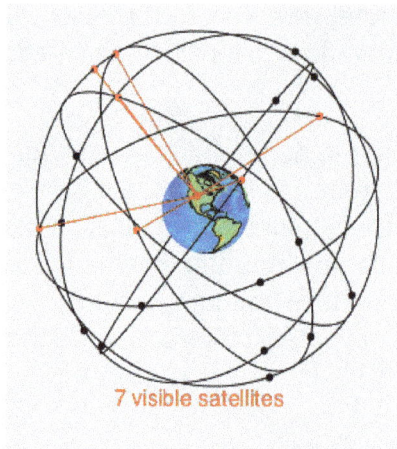

7 visible satellites

A visual example of a 24 satellite GPS constellation in motion with the earth rotating. Notice how the number of *satellites in view* from a given point on the earth's surface, in this example in Golden CO (39.7469° N, 105.2108° W), changes with time.

As of February 2016, there are 32 satellites in the GPS constellation, 31 of which are in use. The additional satellites improve the precision of GPS receiver calculations by providing redundant measurements. With the increased number of satellites, the constellation was changed to a non-uniform arrangement. Such an arrangement was shown to improve reliability and availability of the system, relative to a uniform system, when multiple satellites fail. About nine satellites are vis-ible from any point on the ground at any one time, ensuring considerable redundancy over the minimum four satellites needed for a position.

Control Segment

Ground monitor station used from 1984 to 2007, on display at the Air Force Space & Missile Museum.

The control segment is composed of:

1. a master control station (MCS),

2. an alternate master control station,

3. four dedicated ground antennas, and

4. six dedicated monitor stations.

The MCS can also access U.S. Air Force Satellite Control Network (AFSCN) ground antennas (for additional command and control capability) and NGA (National Geospatial-Intelligence Agency) monitor stations. The flight paths of the satellites are tracked by dedicated U.S. Air Force moni-toring stations in Hawaii, Kwajalein Atoll, Ascension Island, Diego Garcia, Colorado Springs, Colorado and Cape Canaveral, along with shared NGA monitor stations operated in England, Argentina, Ecuador, Bahrain, Australia and Washington DC. The tracking information is sent to the Air Force Space Command MCS at Schriever Air Force Base 25 km (16 mi) ESE of Colorado Springs, which is operated by the 2nd Space Operations Squadron (2 SOPS) of the U.S. Air Force. Then 2 SOPS contacts each GPS satellite regularly with a navigational update using dedicated or shared (AFSCN) ground antennas (GPS dedicated ground antennas are located at Kwajalein, Ascension Island, Diego Garcia, and Cape Canaveral). These updates synchronize the atomic clocks on board the satellites to within a few nanoseconds of each other, and adjust the ephemeris of each satellite's internal orbital model. The updates are created by a Kalman filter that uses inputs from the ground monitoring stations, space weather information, and various other inputs.

Satellite maneuvers are not precise by GPS standards—so to change a satellite's orbit, the satellite must be marked *unhealthy*, so receivers don't use it. After the satellite maneuver, engineers track the new orbit from the ground, upload the new ephemeris, and mark the satellite healthy again.

The Operation Control Segment (OCS) currently serves as the control segment of record. It provides the operational capability that supports GPS users and keeps the GPS system operational and performing within specification.

OCS successfully replaced the legacy 1970s-era mainframe computer at Schriever Air Force Base in September 2007. After installation, the system helped enable upgrades and provide a foundation for a new security architecture that supported U.S. armed forces. OCS will continue to be the ground control system of record until the new segment, Next Generation GPS Operation Control System (OCX), is fully developed and functional.

The new capabilities provided by OCX will be the cornerstone for revolutionizing GPS's mission capabilities, and enabling Air Force Space Command to greatly enhance GPS operational services to U.S. combat forces, civil partners and myriad domestic and international users.

The GPS OCX program also will reduce cost, schedule and technical risk. It is designed to provide 50% sustainment cost savings through efficient software architecture and Performance-Based Logistics. In addition, GPS OCX is expected to cost millions less than the cost to upgrade OCS while providing four times the capability.

The GPS OCX program represents a critical part of GPS modernization and provides significant information assurance improvements over the current GPS OCS program.

- OCX will have the ability to control and manage GPS legacy satellites as well as the next generation of GPS III satellites, while enabling the full array of military signals.

- Built on a flexible architecture that can rapidly adapt to the changing needs of today's and future GPS users allowing immediate access to GPS data and constellation status through secure, accurate and reliable information.

- Provides the warfighter with more secure, actionable and predictive information to enhance situational awareness.

- Enables new modernized signals (L1C, L2C, and L5) and has M-code capability, which the legacy system is unable to do.

- Provides significant information assurance improvements over the current program including detecting and preventing cyber attacks, while isolating, containing and operating during such attacks.

- Supports higher volume near real-time command and control capabilities and abilities.

On September 14, 2011, the U.S. Air Force announced the completion of GPS OCX Preliminary Design Review and confirmed that the OCX program is ready for the next phase of development.

The GPS OCX program has missed major milestones and is pushing the GPS IIIA launch beyond April 2016.

User Segment

The user segment is composed of hundreds of thousands of U.S. and allied military users of the secure GPS Precise Positioning Service, and tens of millions of civil, commercial and scientific users of the Standard Positioning Service. In general, GPS receivers are composed of an antenna, tuned to the frequencies transmitted by the satellites, receiver-processors, and a highly stable clock (often a crystal oscillator). They may also include a display for providing location and speed information to the user. A receiver is often described by its number of channels: this signifies how many satellites it can monitor simultaneously. Originally limited to four or five, this has progressively increased over the years so that, as of 2007, receivers typically have between 12 and 20 channels.

GPS receivers come in a variety of formats, from devices integrated into cars, phones, and watches, to dedicated devices such as these.

GPS receivers may include an input for differential corrections, using the RTCM SC-104 format. This is typically in the form of an RS-232 port at 4,800 bit/s speed. Data is actually sent at a much lower rate, which limits the accuracy of the signal sent using RTCM. Receivers with internal DGPS receivers can outperform those using external RTCM data. As of 2006, even low-cost units commonly include Wide Area Augmentation System (WAAS) receivers.

A typical OEM GPS receiver module measuring 15×17 mm.

A Typical GPS Receiver With Integrated Antenna.

Many GPS receivers can relay position data to a PC or other device using the NMEA 0183 protocol. Although this protocol is officially defined by the National Marine Electronics Association (NMEA), references to this protocol have been compiled from public records, allowing open source tools like gpsd to read the protocol without violating intellectual property laws. Other proprietary protocols exist as well, such as the SiRF and MTK protocols. Receivers can interface with other devices using methods including a serial connection, USB, or Bluetooth.

Applications

While originally a military project, GPS is considered a *dual-use* technology, meaning it has significant military and civilian applications.

GPS has become a widely deployed and useful tool for commerce, scientific uses, tracking, and surveillance. GPS's accurate time facilitates everyday activities such as banking, mobile phone operations, and even the control of power grids by allowing well synchronized hand-off switching.

Civilian

This antenna is mounted on the roof of a hut containing a scientific experiment needing precise timing.

Many civilian applications use one or more of GPS's three basic components: absolute location, relative movement, and time transfer.

- Astronomy: both positional and clock synchronization data is used in astrometry and celestial mechanics calculations. It is also used in amateur astronomy using small telescopes to professionals observatories, for example, while finding extrasolar planets.

- Automated vehicle: applying location and routes for cars and trucks to function without a human driver.

- Cartography: both civilian and military cartographers use GPS extensively.

- Cellular telephony: clock synchronization enables time transfer, which is critical for synchronizing its spreading codes with other base stations to facilitate inter-cell hand-off and support hybrid GPS/cellular position detection for mobile emergency calls and other applications. The first handsets with integrated GPS launched in the late 1990s. The U.S. Federal Communications Commission (FCC) mandated the feature in either the handset or in the towers (for use in triangulation) in 2002 so emergency services could locate 911 callers. Third-party software developers later gained access to GPS APIs from Nextel upon launch, followed by Sprint in 2006, and Verizon soon thereafter.

- Clock synchronization: the accuracy of GPS time signals (±10 ns) is second only to the atomic clocks they are based on.

- Disaster relief/emergency services: depend upon GPS for location and timing capabilities.

- GPS-equipped radiosondes and dropsondes: measure and calculate the atmospheric pressure, wind speed and direction up to 27 km from the Earth's surface

- Radio occultation for weather and atmospheric science applications.

- Fleet tracking: the use of GPS technology to identify, locate and maintain contact reports with one or more fleet vehicles in real-time.

- Geofencing: vehicle tracking systems, person tracking systems, and pet tracking systems use GPS to locate a vehicle, person, or pet. These devices are attached to the vehicle, person, or the pet collar. The application provides continuous tracking and mobile or Internet updates should the target leave a designated area.

- Geotagging: applying location coordinates to digital objects such as photographs (in Exif data) and other documents for purposes such as creating map overlays with devices like Nikon GP-1

- GPS aircraft tracking

- GPS for mining: the use of RTK GPS has significantly improved several mining operations such as drilling, shoveling, vehicle tracking, and surveying. RTK GPS provides centimeter-level positioning accuracy.

- GPS data mining: It is possible to use GPS data from multiple users to understand movement patterns. It is possible to aggregate data from multiple users to understand common trajectories and interesting locations.

- GPS tours: location determines what content to display; for instance, information about an approaching point of interest.

- Navigation: navigators value digitally precise velocity and orientation measurements.

- Phasor measurements: GPS enables highly accurate timestamping of power system measurements, making it possible to compute phasors.

- Recreation: for example, geocaching, geodashing, GPS drawing and waymarking.

- Robotics: self-navigating, autonomous robots using a GPS sensors, which calculate latitude, longitude, time, speed, and heading.

- Sport: used in football and rugby for the control and analysis of the training load.

- Surveying: surveyors use absolute locations to make maps and determine property boundaries.

- Tectonics: GPS enables direct fault motion measurement of earthquakes. Between earthquakes GPS can be used to measure crustal motion and deformation to estimate seismic strain buildup for creating seismic hazard maps.

- Telematics: GPS technology integrated with computers and mobile communications technology in automotive navigation systems

Restrictions on Civilian Use

The U.S. government controls the export of some civilian receivers. All GPS receivers capable of functioning above 18 km (60,000 feet) altitude and 515 m/s (1,000 knots), or designed or modified for use with unmanned air vehicles like, e.g., ballistic or cruise missile systems, are classified as munitions (weapons)—which means they require State Department export licenses.

This rule applies even to otherwise purely civilian units that only receive the L1 frequency and the C/A (Coarse/Acquisition) code.

Disabling operation above these limits exempts the receiver from classification as a munition. Vendor interpretations differ. The rule refers to operation at both the target altitude and speed, but some receivers stop operating even when stationary. This has caused problems with some amateur radio balloon launches that regularly reach 30 km (100,000 feet).

These limits only apply to units or components exported from the USA. A growing trade in various components exists, including GPS units from other countries. These are expressly sold as ITAR-free.

Military

Attaching a GPS guidance kit to a dumb bomb, March 2003.

M982 Excalibur GPS-guided artillery shell.

As of 2009, military GPS applications include:

- Navigation: Soldiers use GPS to find objectives, even in the dark or in unfamiliar territory, and to coordinate troop and supply movement. In the United States armed forces, commanders use the *Commanders Digital Assistant* and lower ranks use the *Soldier Digital Assistant*.

- Target tracking: Various military weapons systems use GPS to track potential ground and air targets before flagging them as hostile. These weapon systems pass target coordinates to precision-guided munitions to allow them to engage targets accurately. Military aircraft, particularly in air-to-ground roles, use GPS to find targets.

- Missile and projectile guidance: GPS allows accurate targeting of various military weapons including ICBMs, cruise missiles, precision-guided munitions and Artillery projectiles. Embedded GPS receivers able to withstand accelerations of 12,000 g or about 118 km/s^2 have been developed for use in 155-millimeter (6.1 in) howitzers.

- Search and rescue.

- Reconnaissance: Patrol movement can be managed more closely.

- GPS satellites carry a set of nuclear detonation detectors consisting of an optical sensor (Y-sensor), an X-ray sensor, a dosimeter, and an electromagnetic pulse (EMP) sensor (W-sensor), that form a major portion of the United States Nuclear Detonation Detection System. General William Shelton has stated that future satellites may drop this feature to save money.

GPS type navigation was first used in war in the 1991 Persian Gulf War, before GPS was fully developed in 1995, to assist Coalition Forces to navigate and perform maneuvers in the war. The war also demonstrated the vulnerability of GPS to being jammed, when Iraqi forces added noise to the weak GPS signal transmission to protect Iraqi targets.

Communication

The navigational signals transmitted by GPS satellites encode a variety of information including satellite positions, the state of the internal clocks, and the health of the network. These signals are transmitted on two separate carrier frequencies that are common to all satellites in the network. Two different encodings are used: a public encoding that enables lower resolution navigation, and an encrypted encoding used by the U.S. military.

Message Format

GPS message format	
Subframes	**Description**
1	Satellite clock, GPS time relationship
2–3	Ephemeris (precise satellite orbit)
4–5	Almanac component (satellite network synopsis, error correction)

Each GPS satellite continuously broadcasts a *navigation message* on L1 C/A and L2 P/Y frequencies at a rate of 50 bits per second. Each complete message takes 750 seconds (12 1/2 min-utes) to complete. The message structure has a basic format of a 1500-bit-long frame made up of five subframes, each subframe being 300 bits (6 seconds) long. Subframes 4 and 5 are subcommutated 25 times each, so that a complete data message requires the transmission of 25 full frames. Each subframe consists of ten words, each 30 bits long. Thus, with 300 bits in a subframe times 5 subframes in a frame times 25 frames in a message, each message is 37,500 bits long. At a transmission rate of 50-bit/s, this gives 750 seconds to transmit an entire almanac message (GPS). Each 30-second frame begins precisely on the minute or half-minute as indicated by the atomic clock on each satellite.

The first subframe of each frame encodes the week number and the time within the week, as well as the data about the health of the satellite. The second and the third subframes contain the *ephemeris* – the precise orbit for the satellite. The fourth and fifth subframes contain the *almanac*, which contains coarse orbit and status information for up to 32 satellites in the constellation as well as data related to error correction. Thus, to obtain an accurate satellite location from this transmitted message, the receiver must demodulate the message from each satellite it includes in its solution for 18 to 30 seconds. To collect all transmitted almanacs, the receiver must demodulate the message for 732 to 750 seconds or 12 1/2 minutes.

All satellites broadcast at the same frequencies, encoding signals using unique code division mul-tiple access (CDMA) so receivers can distinguish individual satellites from each other. The system uses two distinct CDMA encoding types: the coarse/acquisition (C/A) code, which is accessible by the general public, and the precise (P(Y)) code, which is encrypted so that only the U.S. military and other NATO nations who have been given access to the encryption code can access it.

The ephemeris is updated every 2 hours and is generally valid for 4 hours, with provisions for up-dates every 6 hours or longer in non-nominal conditions. The almanac is updated typically every 24 hours. Additionally, data for a few weeks following is uploaded in case of transmission updates that delay data upload.

Satellite Frequencies

GPS frequency overview		
Band	**Frequency**	**Description**
L1	1575.42 MHz	Coarse-acquisition (C/A) and encrypted precision (P(Y)) codes, plus the L1 civilian (L1C) and military (M) codes on future Block III satellites.
L2	1227.60 MHz	P(Y) code, plus the L2C and military codes on the Block IIR-M and newer satellites.
L3	1381.05 MHz	Used for nuclear detonation (NUDET) detection.
L4	1379.913 MHz	Being studied for additional ionospheric correction.
L5	1176.45 MHz	Proposed for use as a civilian safety-of-life (SoL) signal.

All satellites broadcast at the same two frequencies, 1.57542 GHz (L1 signal) and 1.2276 GHz (L2 signal). The satellite network uses a CDMA spread-spectrum technique where the low-bitrate message data is encoded with a high-rate pseudo-random (PRN) sequence that is different for each satellite. The receiver must be aware of the PRN codes for each satellite to reconstruct the actual message data. The C/A code, for civilian use, transmits data at 1.023 million chips per second, whereas the P code, for U.S. military use, transmits at 10.23 million chips per second. The actual internal reference of the satellites is 10.22999999543 MHz to compensate for relativistic effects that make observers on the Earth perceive a different time reference with respect to the transmitters in orbit. The L1 carrier is modulated by both the C/A and P codes, while the L2 carrier is only modulated by the P code. The P code can be encrypted as a so-called P(Y) code that is only available to military equipment with a proper decryption key. Both the C/A and P(Y) codes impart the precise time-of-day to the user.

The L3 signal at a frequency of 1.38105 GHz is used to transmit data from the satellites to ground stations. This data is used by the United States Nuclear Detonation (NUDET) Detection System (USNDS) to detect, locate, and report nuclear detonations (NUDETs) in the Earth's atmosphere and near space. One usage is the enforcement of nuclear test ban treaties.

The L4 band at 1.379913 GHz is being studied for additional ionospheric correction.

The L5 frequency band at 1.17645 GHz was added in the process of GPS modernization. This frequency falls into an internationally protected range for aeronautical navigation, promising little or no interference under all circumstances. The first Block IIF satellite that provides this signal was launched in 2010. The L5 consists of two carrier components that are in phase quadrature with each other. Each carrier component is bi-phase shift key (BPSK) modulated by a separate bit train. "L5, the third civil GPS signal, will eventually support safety-of-life applications for aviation and provide improved availability and accuracy."

A conditional waiver has recently (2011-01-26) been granted to LightSquared to operate a terrestrial broadband service near the L1 band. Although LightSquared had applied for a license to operate in the 1525 to 1559 band as early as 2003 and it was put out for public comment, the FCC asked LightSquared to form a study group with the GPS community to test GPS receivers and identify issue that might arise due to the larger signal power from the LightSquared terrestrial network. The GPS community had not objected to the LightSquared (formerly MSV and SkyTerra) applications until November 2010, when LightSquared applied for a modification to its Ancillary Terrestrial Component (ATC) authorization. This filing (SAT-MOD-20101118-00239) amounted to a request to run several orders of magnitude more power in the same frequency band for terrestrial base stations, essentially repurposing what was supposed to be a "quiet neighborhood" for signals from space as the equivalent of a cellular network. Testing in the first half of 2011 has demonstrated that the impact of the lower 10 MHz of spectrum is minimal to GPS devices (less than 1% of the total GPS devices are affected). The upper 10 MHz intended for use by LightSquared may have some impact on GPS devices. There is some concern that this may seriously degrade the GPS signal for many consumer uses. Aviation Week magazine reports that the latest testing (June 2011) confirms "significant jamming" of GPS by LightSquared's system.

Demodulation and Decoding

Because all of the satellite signals are modulated onto the same L1 carrier frequency, the signals must be separated after demodulation. This is done by assigning each satellite a unique binary sequence known as a Gold code. The signals are decoded after demodulation using addition of the Gold codes corresponding to the satellites monitored by the receiver.

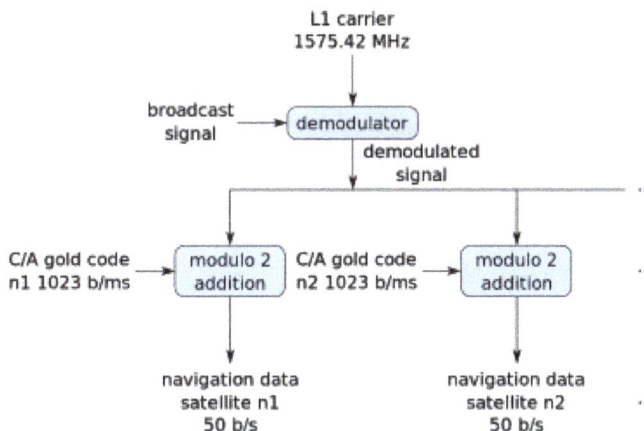

Demodulating and Decoding GPS Satellite Signals using the Coarse/Acquisition Gold code.

If the almanac information has previously been acquired, the receiver picks the satellites to listen for by their PRNs, unique numbers in the range 1 through 32. If the almanac information is not in memory, the receiver enters a search mode until a lock is obtained on one of the satellites. To obtain a lock, it is necessary that there be an unobstructed line of sight from the receiver to the satellite. The receiver can then acquire the almanac and determine the satellites it should listen for. As it detects each satellite's signal, it identifies it by its distinct C/A code pattern. There can be a delay of up to 30 seconds before the first estimate of position because of the need to read the ephemeris data.

Navigation Equations

Problem Description

The receiver uses messages received from satellites to determine the satellite positions and time sent. The x, y, and z components of satellite position and the time sent are designated as $[x_i, y_i, z_i, s_i]$ where the subscript i denotes the satellite and has the value 1, 2, ..., n, where $n \geq 4$. When the time of message reception indicated by the on-board receiver clock is $\tilde{t}_i$, the true reception time is $t_i = \tilde{t}_i - b$, where b is the receiver's clock bias from the much more accurate GPS system clocks employed by the satellites. The receiver clock bias is the same for all received satellite signals (assuming the satellite clocks are all perfectly synchronized). The message's transit time is $\tilde{t}_i - b - s_i$, where s_i is the satellite time. Assuming the message traveled at the speed of light, c, the distance traveled is $(\tilde{t}_i - b - s_i)\,c$.

For n satellites, the equations to satisfy are:

$$(x - x_i)^2 + (y - y_i)^2 + (z - z_i)^2 = \left([\tilde{t}_i - b - s_i]c\right)^2, i = 1, 2, \ldots, n$$

or in terms of *pseudoranges*, $p_i = \left(\tilde{t}_i - s_i\right)c$, as

$$\sqrt{(x - x_i)^2 + (y - y_i)^2 + (z - z_i)^2} + bc = p_i, i = 1, 2, \ldots, n.$$

Since the equations have four unknowns [x, y, z, b]—the three components of GPS receiver position and the clock bias—signals from at least four satellites are necessary to attempt solving these equations. They can be solved by algebraic or numerical methods. Existence and uniqueness of GPS solutions are discussed by Abell and Chaffee. When n is greater than 4 this system is overdetermined and a fitting method must be used.

With each combination of satellites, GDOP quantities can be calculated based on the relative sky directions of the satellites used. The receiver location is expressed in a specific coordinate system, such as latitude and longitude using the WGS 84 geodetic datum or a country-specific system.

Geometric Interpretation

The GPS equations can be solved by numerical and analytical methods. Geometrical interpretations can enhance the understanding of these solution methods.

Spheres

The measured ranges, called pseudoranges, contain clock errors. In a simplified idealization in which the ranges are synchronized, these true ranges represent the radii of spheres, each centered on one of the transmitting satellites. The solution for the position of the receiver is then at the intersection of the surfaces of three of these spheres. If more than the minimum number of ranges is available, a near intersection of more than three sphere surfaces could be found via, e.g. least squares.

Hyperboloids

If the distance traveled between the receiver and satellite i and the distance traveled between the receiver and satellite j are subtracted, the result is $(\tilde{t}_i - s_i)\,c - (\tilde{t}_j - s_j)\,c$, which only involves known or measured quantities. The locus of points having a constant difference in distance to two points (here, two satellites) is a hyperboloid. Thus, from four or more measured reception times, the receiver can be placed at the intersection of the surfaces of three or more hyperboloids.

Spherical Cones

The solution space [x, y, z, b] can be seen as a four-dimensional geometric space. In that case each of the equations describes a spherical cone, with the cusp located at the satellite, and the base a sphere around the satellite. The receiver is at the intersection of four or more of such cones.

Solution Methods

Least Squares

When more than four satellites are available, the calculation can use the four best, or more than four simultaneously (up to all visible satellites), depending on the number of receiver channels, processing capability, and geometric dilution of precision (GDOP).

Using more than four involves an over-determined system of equations with no unique solution; such a system can be solved by a least-squares or weighted least squares method.

$$\left(\hat{x}, \hat{y}, \hat{z}, \hat{b}\right) = \underset{(x,y,z,b)}{\arg\min} \sum_i \left(\sqrt{(x-x_i)^2 + (y-y_i)^2 + (z-z_i)^2} + bc - p_i \right)^2$$

Iterative

Both the equations for four satellites, or the least squares equations for more than four, are non-linear and need special solution methods. A common approach is by iteration on a linearized form of the equations, (e.g., Gauss–Newton algorithm).

The GPS system was initially developed assuming use of a numerical least-squares solution method—i.e., before closed-form solutions were found.

Closed-form

One closed-form solution to the above set of equations was developed by S. Bancroft. Its properties are well known; in particular, proponents claim it is superior in low-GDOP situations, compared to iterative least squares methods.

Bancroft's method is algebraic, as opposed to numerical, and can be used for four or more satellites. When four satellites are used, the key steps are inversion of a 4x4 matrix and solution of a single-variable quadratic equation. Bancroft's method provides one or two solutions for the unknown quantities. When there are two (usually the case), only one is a near-Earth sensible solution.

When a receiver uses more than four satellites for a solution, Bancroft uses the generalized inverse (i.e., the pseudoinverse) to find a solution. However, a case has been made that iterative methods (e.g., Gauss–Newton algorithm) for solving over-determined non-linear least squares (NLLS) problems generally provide more accurate solutions.

Leick et al. (2015) states that "Bancroft's (1985) solution is a very early, if not the first, closed-form solution." Other closed-form solutions were published afterwards, although their adoption in practice is unclear.

Error Sources and Analysis

GPS error analysis examines error sources in GPS results and the expected size of those errors. GPS makes corrections for receiver clock errors and other effects, but some residual errors remain uncorrected. Error sources include signal arrival time measurements, numerical calculations, atmospheric effects (ionospheric/tropospheric delays), ephemeris and clock data, multipath signals, and natural and artificial interference. Magnitude of residual errors from these sources depends on geometric dilution of precision. Artificial errors may result from jamming devices and threaten ships and aircraft or from intentional signal degradation through selective availability, which limited accuracy to ~6–12 m, but has been switched off since May 1, 2000.

Accuracy Enhancement and Surveying

Augmentation

Integrating external information into the calculation process can materially improve accuracy. Such augmentation systems are generally named or described based on how the information arrives. Some systems transmit additional error information (such as clock drift, ephemera, or ionospheric delay), others characterize prior errors, while a third group provides additional navigational or vehicle information.

Examples of augmentation systems include the Wide Area Augmentation System (WAAS), European Geostationary Navigation Overlay Service (EGNOS), Differential GPS (DGPS), inertial navigation systems (INS) and Assisted GPS. The standard accuracy of about 15 meters (49 feet) can be augmented to 3–5 meters (9.8–16.4 ft) with DGPS, and to about 3 meters (9.8 feet) with WAAS.

Precise Monitoring

Accuracy can be improved through precise monitoring and measurement of existing GPS signals in additional or alternate ways.

The largest remaining error is usually the unpredictable delay through the ionosphere. The spacecraft broadcast ionospheric model parameters, but some errors remain. This is one reason GPS spacecraft transmit on at least two frequencies, L1 and L2. Ionospheric delay is a well-defined function of frequency and the total electron content (TEC) along the path, so measuring the arrival time difference between the frequencies determines TEC and thus the precise ionospheric delay at each frequency.

Military receivers can decode the P(Y) code transmitted on both L1 and L2. Without decryption keys, it is still possible to use a *codeless* technique to compare the P(Y) codes on L1 and L2 to gain much of the same error information. However, this technique is slow, so it is currently available only on specialized surveying equipment. In the future, additional civilian codes are expected to be transmitted on the L2 and L5 frequencies. All users will then be able to perform dual-frequency measurements and directly compute ionospheric delay errors.

A second form of precise monitoring is called *Carrier-Phase Enhancement* (CPGPS). This corrects the error that arises because the pulse transition of the PRN is not instantaneous, and thus the correlation (satellite–receiver sequence matching) operation is imperfect. CPGPS uses the L1 carrier wave, which has a period of $\frac{1s}{1575.42 \times 10^6} = 0.63475\text{ns} \approx 1\text{ns}$, which is about one-thousandth of

the C/A Gold code bit period of $\frac{1s}{1023 \times 10^3} = 977.5\text{ns} \approx 1000\text{ns}$, to act as an additional clock signal and resolve the uncertainty. The phase difference error in the normal GPS amounts to 2–3 meters (7–10 ft) of ambiguity. CPGPS working to within 1% of perfect transition reduces this error to 3 centimeters (1.2 in) of ambiguity. By eliminating this error source, CPGPS coupled with DGPS normally realizes between 20–30 centimeters (8–12 in) of absolute accuracy.

Relative Kinematic Positioning (RKP) is a third alternative for a precise GPS-based positioning system. In this approach, determination of range signal can be resolved to a precision of less than 10 centimeters (4 in). This is done by resolving the number of cycles that the signal is transmitted and received by the receiver by using a combination of differential GPS (DGPS) correction data, transmitting GPS signal phase information and ambiguity resolution techniques via statistical tests—possibly with processing in real-time (real-time kinematic positioning, RTK).

Timekeeping

Leap Seconds

While most clocks derive their time from Coordinated Universal Time (UTC), the atomic clocks on the satellites are set to GPS time. The dif-ference is that GPS time is not corrected to match the rotation of the Earth, so it does not contain leap seconds or other corrections that are periodically added to UTC. GPS time was set to match UTC in 1980, but has since diverged. The lack of corrections means that GPS time remains at a constant offset with International Atomic Time (TAI) (TAI – GPS = 19 seconds). Periodic correc-tions are performed to the on-board clocks to keep them synchronized with ground clocks.

The GPS navigation message includes the difference between GPS time and UTC. As of July 2015, GPS time is 17 seconds ahead of UTC because of the leap second added to UTC on June 30, 2015. Receivers subtract this offset from GPS time to calculate UTC and specific timezone values. New GPS units may not show the correct UTC time until after receiving the UTC offset message. The GPS-UTC offset field can accommodate 255 leap seconds (eight bits).

Accuracy

GPS time is theoretically accurate to about 14 nanoseconds. However, most receivers lose accuracy in the interpretation of the signals and are only accurate to 100 nanoseconds.

Format

As opposed to the year, month, and day format of the Gregorian calendar, the GPS date is ex-pressed as a week number and a seconds-into-week number. The week number is transmitted as a ten-bit field in the C/A and P(Y) navigation messages, and so it becomes zero again every 1,024 weeks (19.6 years). GPS week zero started at 00:00:00 UTC (00:00:19 TAI) on January 6, 1980, and the week number became zero again for the first time at 23:59:47 UTC on August 21,

1999 (00:00:19 TAI on August 22, 1999). To determine the current Gregorian date, a GPS receiver must be provided with the approximate date (to within 3,584 days) to correctly translate the GPS date signal. To address this concern the modernized GPS navigation message uses a 13-bit field that only repeats every 8,192 weeks (157 years), thus lasting until the year 2137 (157 years after GPS week zero).

Carrier Phase Tracking (Surveying)

Another method that is used in surveying applications is carrier phase tracking. The period of the carrier frequency multiplied by the speed of light gives the wavelength, which is about 0.19 meters for the L1 carrier. Accuracy within 1% of wavelength in detecting the leading edge reduces this component of pseudorange error to as little as 2 millimeters. This compares to 3 meters for the C/A code and 0.3 meters for the P code.

However, 2 millimeter accuracy requires measuring the total phase—the number of waves multiplied by the wavelength plus the fractional wavelength, which requires specially equipped receivers. This method has many surveying applications. It is accurate enough for real-time tracking of the very slow motions of tectonic plates, typically 0–100 mm (0–4 inches) per year.

Triple differencing followed by numerical root finding, and a mathematical technique called least squares can estimate the position of one receiver given the position of another. First, compute the difference between satellites, then between receivers, and finally between epochs. Other orders of taking differences are equally valid. Detailed discussion of the errors is omitted.

The satellite carrier total phase can be measured with ambiguity as to the number of cycles. Let $\phi(r_i, s_j, t_k)$ denote the phase of the carrier of satellite j measured by receiver i at time t_k. This notation shows the meaning of the subscripts i, j, and k. The receiver (r), satellite (s), and time (t) come in alphabetical order as arguments of ϕ and to balance readability and conciseness, let $\phi_{i,j,k} = \phi(r_i, s_j, t_k)$ be a concise abbreviation. Also we define three functions, : $\Delta^r, \Delta^s, \Delta^t$, which return differences between receivers, satellites, and time points, respectively. Each function has variables with three subscripts as its arguments. These three functions are defined below. If $\alpha_{i,j,k}$ is a function of the three integer arguments, i, j, and k then it is a valid argument for the functions, : $\Delta^r, \Delta^s, \Delta^t$, with the values defined as

$$\Delta^r(\alpha_{i,j,k}) = \alpha_{i+1,j,k} - \alpha_{i,j,k},$$

$$\Delta^s(\alpha_{i,j,k}) = \alpha_{i,j+1,k} - \alpha_{i,j,k}, \text{ and}$$

$$\Delta^t(\alpha_{i,j,k}) = \alpha_{i,j,k+1} - \alpha_{i,j,k}.$$

Receiver clock errors can be approximately eliminated by differencing the phases measured from satellite 1 with that from satellite 2 at the same epoch. This difference is designated as $\Delta^s(\phi_{1,1,1}) = \phi_{1,2,1} - \phi_{1,1,1}$

Double differencing computes the difference of receiver 1's satellite difference from that of receiver 2. This approximately eliminates satellite clock errors. This double difference is:

$$\Delta^r(\Delta^s(\phi_{1,1,1})) = \Delta^r(\phi_{1,2,1} - \phi_{1,1,1}) = \Delta^r(\phi_{1,2,1}) - \Delta^r(\phi_{1,1,1}) = (\phi_{2,2,1} - \phi_{1,2,1}) - (\phi_{2,1,1} - \phi_{1,1,1})$$

Triple differencing subtracts the receiver difference from time 1 from that of time 2. This eliminates the ambiguity associated with the integral number of wavelengths in carrier phase provided this ambiguity does not change with time. Thus the triple difference result eliminates practically all clock bias errors and the integer ambiguity. Atmospheric delay and satellite ephemeris errors have been significantly reduced. This triple difference is:

$$\Delta^t(\Delta^r(\Delta^s(\phi_{1,1,1})))$$

Triple difference results can be used to estimate unknown variables. For example, if the position of receiver 1 is known but the position of receiver 2 unknown, it may be possible to estimate the position of receiver 2 using numerical root finding and least squares. Triple difference results for three independent time pairs may be sufficient to solve for receiver 2's three position components. This may require a numerical procedure. An approximation of receiver 2's position is required to use such a numerical method. This initial value can probably be provided from the navigation message and the intersection of sphere surfaces. Such a reasonable estimate can be key to successful multidimensional root finding. Iterating from three time pairs and a fairly good initial value produces one observed triple difference result for receiver 2's position. Processing additional time pairs can improve accuracy, overdetermining the answer with multiple solutions. Least squares can estimate an overdetermined system. Least squares determines the position of receiver 2 that best fits the observed triple difference results for receiver 2 positions under the criterion of minimizing the sum of the squares.

Regulatory Spectrum Issues Concerning GPS Receivers

In the United States, GPS receivers are regulated under the Federal Communications Commission's (FCC) Part 15 rules. As indicated in the manuals of GPS-enabled devices sold in the United States, as a Part 15 device, it "must accept any interference received, including interference that may cause undesired operation." With respect to GPS devices in particular, the FCC states that GPS receiver manufacturers, "must use receivers that reasonably discriminate against reception of signals outside their allocated spectrum." For the last 30 years, GPS receivers have operated next to the Mobile Satellite Service band, and have discriminated against reception of mobile satellite services, such as Inmarsat, without any issue.

The spectrum allocated for GPS L1 use by the FCC is 1559 to 1610 MHz, while the spectrum allocated for satellite-to-ground use owned by Lightsquared is the Mobile Satellite Service band. Since 1996, the FCC has authorized licensed use of the spectrum neighboring the GPS band of 1525 to 1559 MHz to the Virginia company LightSquared. On March 1, 2001, the FCC received an application from LightSquared's predecessor, Motient Services to use their allocated frequencies for an integrated satellite-terrestrial service. In 2002, the U.S. GPS Industry Council came to an out-of-band-emissions (OOBE) agreement with LightSquared to prevent transmissions from LightSquared's ground-based stations from emitting transmissions into the neighboring GPS band of 1559 to 1610 MHz. In 2004, the FCC adopted the OOBE agreement in its authorization for LightSquared to deploy a ground-based network ancillary to their satellite system – known as the Ancillary Tower Components (ATCs) – "We will authorize MSS ATC subject to conditions that ensure that the added terrestrial component remains ancillary to the principal MSS offering. We do not intend, nor will we permit, the terrestrial component to become a stand-alone service." This authorization was reviewed and approved by the U.S. Interdepartment Radio Advisory Committee, which includes the U.S. Department of Agriculture, U.S. Air Force, U.S. Army, U.S. Coast Guard, Federal Aviation Administration, National Aeronautics and Space Administration, Interior, and U.S. Department of Transportation.

In January 2011, the FCC conditionally authorized LightSquared's wholesale customers—such as Best Buy, Sharp, and C Spire—to only purchase an integrated satellite-ground-based service from LightSquared and re-sell that integrated service on devices that are equipped to only use the ground-based signal using LightSquared's allocated frequencies of 1525 to 1559 MHz. In December 2010, GPS receiver manufacturers expressed concerns to the FCC that LightSquared's signal would interfere with GPS receiver devices although the FCC's policy considerations leading up to the January 2011 order did not pertain to any proposed changes to the maximum number of ground-based LightSquared stations or the maximum power at which these stations could operate. The January 2011 order makes final authorization contingent upon studies of GPS interference issues carried out by a LightSquared led working group along with GPS industry and Federal agency participation. On February 14, 2012, the FCC initiated proceedings to vacate LightSquared's Conditional Waiver Order based on the NTIA's conclusion that there was currently no practical way to mitigate potential GPS interference.

GPS receiver manufacturers design GPS receivers to use spectrum beyond the GPS-allocated band. In some cases, GPS receivers are designed to use up to 400 MHz of spectrum in either direction of the L1 frequency of 1575.42 MHz, because mobile satellite services in those regions are broadcasting from space to ground, and at power levels commensurate with mobile satellite services. However, as regulated under the FCC's Part 15 rules, GPS receivers are not warranted protection from signals outside GPS-allocated spectrum. This is why GPS operates next to the Mobile Satellite Service band, and also why the Mobile Satellite Service band operates next to GPS. The symbiotic relationship of spectrum allocation ensures that users of both bands are able to operate cooperatively and freely.

The FCC adopted rules in February 2003 that allowed Mobile Satellite Service (MSS) licensees such as LightSquared to construct a small number of ancillary ground-based towers in their licensed spectrum to "promote more efficient use of terrestrial wireless spectrum." In those 2003 rules, the FCC stated "As a preliminary matter, terrestrial [Commercial Mobile Radio Service ("CMRS")] and MSS ATC are expected to have different prices, coverage, product acceptance and distribution; therefore, the two services appear, at best, to be imperfect substitutes for one another that would be operating in predominately different market segments... MSS ATC is unlikely to compete directly with terrestrial CMRS for the same customer base...". In 2004, the FCC clarified that the ground-based towers would be ancillary, noting that "We will authorize MSS ATC subject to conditions that ensure that the added terrestrial component remains ancillary to the principal MSS offering. We do not intend, nor will we permit, the terrestrial component to become a stand-alone service." In July 2010, the FCC stated that it expected LightSquared to use its authority to offer an integrated satellite-terrestrial service to "provide mobile broadband services similar to those provided by terrestrial mobile providers and enhance competition in the mobile broadband sector." However, GPS receiver manufacturers have argued that LightSquared's licensed spectrum of 1525 to 1559 MHz was never envisioned as being used for high-speed wireless broadband based on the 2003 and 2004 FCC ATC rulings making clear that the Ancillary Tower Component (ATC) would be, in fact, ancillary to the primary satellite component. To build public support of efforts to continue the 2004 FCC authorization of LightSquared's ancillary terrestrial component vs. a simple ground-based LTE service in the Mobile Satellite Service band, GPS receiver manufacturer Trimble Navigation Ltd. formed the "Coalition To Save Our GPS."

The FCC and LightSquared have each made public commitments to solve the GPS interference issue before the network is allowed to operate. However, according to Chris Dancy of the Aircraft Owners and Pilots Association, airline pilots with the type of systems that would be affected "may go off course and not even realize it." The problems could also affect the Federal Aviation Administration upgrade to the air traffic control system, United States Defense Department guidance, and local emergency services including 911.

On February 14, 2012, the U.S. Federal Communications Commission (FCC) moved to bar LightSquared's planned national broadband network after being informed by the National Telecommunications and Information Administration (NTIA), the federal agency that coordinates spectrum uses for the military and other federal government entities, that "there is no practical way to mitigate potential interference at this time". LightSquared is challenging the FCC's action.

Other Systems

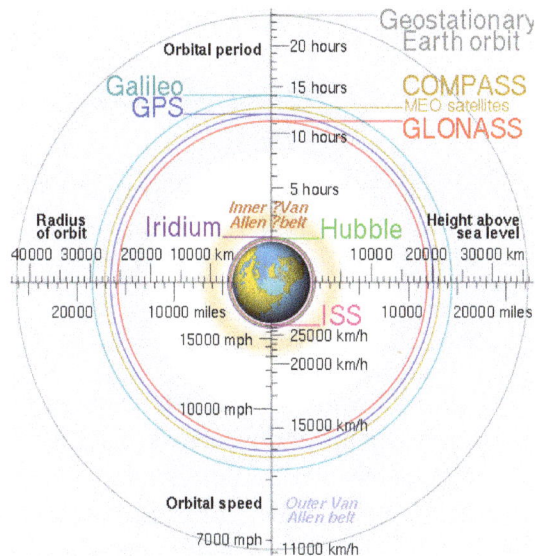

Comparison of geostationary, GPS, GLONASS, Galileo, Compass (MEO), International Space Station, Hubble Space Telescope and Iridium constellation orbits, with the Van Allen radiation belts and the Earth to scale. The Moon's orbit is around 9 times larger than geostationary orbit. (In the SVG file, hover over an orbit or its label to highlight it; click to load its article.)

Other satellite navigation systems in use or various states of development include:

- GLONASS – Russia's global navigation system. Fully operational worldwide.

- Galileo – a global system being developed by the European Union and other partner countries, planned to be operational by 2016 (and fully deployed by 2020)

- Beidou – People's Republic of China's regional system, currently limited to Asia and the West Pacific, global coverage planned to be operational by 2020

- IRNSS (NAVIC) – India's regional navigation system, covering India and Northern Indian Ocean

- QZSS – Japanese regional system covering Asia and Oceania

Wireless Sensor Network

Wireless sensor networks (WSN), sometimes called wireless sensor and actuator networks (WSAN), are spatially distributed autonomous sensors to *monitor* physical or environmental conditions, such as temperature, sound, pressure, etc. and to cooperatively pass their data through the network to a main location. The more modern networks are bi-directional, also enabling *control* of sensor activity. The development of wireless sensor networks was motivated by military applications such as battlefield surveillance; today such networks are used in many industrial and consumer applications, such as industrial process monitoring and control, machine health monitoring, and so on.

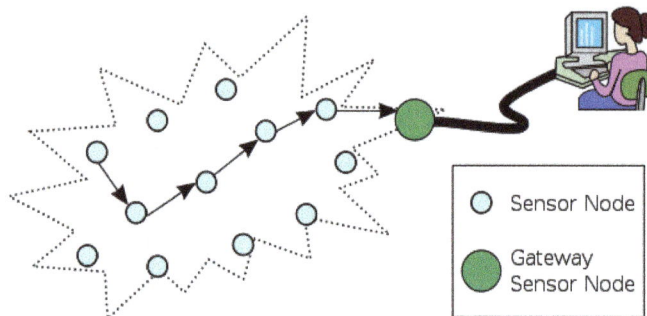

Typical multi-hop wireless sensor network architecture

The WSN is built of "nodes" – from a few to several hundreds or even thousands, where each node is connected to one (or sometimes several) sensors. Each such sensor network node has typically several parts: a radio transceiver with an internal antenna or connection to an external antenna, a microcontroller, an electronic circuit for interfacing with the sensors and an energy source, usually a battery or an embedded form of energy harvesting. A sensor node might vary in size from that of a shoebox down to the size of a grain of dust, although functioning "motes" of genuine microscopic dimensions have yet to be created. The cost of sensor nodes is similarly variable, ranging from a few to hundreds of dollars, depending on the complexity of the individual sensor nodes. Size and cost constraints on sensor nodes result in corresponding constraints on resources such as energy, memory, computational speed and communications bandwidth. The topology of the WSNs can vary from a simple star network to an advanced multi-hop wireless mesh network. The propagation technique between the hops of the network can be routing or flooding.

In computer science and telecommunications, wireless sensor networks are an active research area with numerous workshops and conferences arranged each year, for example IPSN, SenSys, and EWSN.

Application

Area Monitoring

Area monitoring is a common application of WSNs. In area monitoring, the WSN is deployed over a region where some phenomenon is to be monitored. A military example is the use of sensors detect enemy intrusion; a civilian example is the geo-fencing of gas or oil pipelines.

Health Care Monitoring

The medical applications can be of two types: wearable and implanted. Wearable devices are used on the body surface of a human or just at close proximity of the user. The implantable medical devices are those that are inserted inside human body. There are many other applications too e.g. body position measurement and location of the person, overall monitoring of ill patients in hospitals and at homes. Body-area networks can collect information about an individual's health, fitness, and energy expenditure.

Environmental/Earth Sensing

There are many applications in monitoring environmental parameters, examples of which are given below. They share the extra challenges of harsh environments and reduced power supply.

Air Pollution Monitoring

Wireless sensor networks have been deployed in several cities (Stockholm, London, and Brisbane) to monitor the concentration of dangerous gases for citizens. These can take advantage of the ad hoc wireless links rather than wired installations, which also make them more mobile for testing readings in different areas.

Forest Fire Detection

A network of Sensor Nodes can be installed in a forest to detect when a fire has started. The nodes can be equipped with sensors to measure temperature, humidity and gases which are produced by fire in the trees or vegetation. The early detection is crucial for a successful action of the firefighters; thanks to Wireless Sensor Networks, the fire brigade will be able to know when a fire is started and how it is spreading.

Landslide Detection

A landslide detection system makes use of a wireless sensor network to detect the slight movements of soil and changes in various parameters that may occur before or during a landslide. Through the data gathered it may be possible to know the impending occurrence of landslides long before it actually happens.

Water Quality Monitoring

Water quality monitoring involves analyzing water properties in dams, rivers, lakes & oceans, as well as underground water reserves. The use of many wireless distributed sensors enables the creation of a more accurate map of the water status, and allows the permanent deployment of monitoring stations in locations of difficult access, without the need of manual data retrieval.

Natural Disaster Prevention

Wireless sensor networks can effectively act to prevent the consequences of natural disasters, like floods. Wireless nodes have successfully been deployed in rivers where changes of the water levels have to be monitored in real time.

Industrial Monitoring

Machine Health Monitoring

Wireless sensor networks have been developed for machinery condition-based maintenance (CBM) as they offer significant cost savings and enable new functionality.

Wireless sensors can be placed in locations difficult or impossible to reach with a wired system, such as rotating machinery and untethered vehicles.

Data Logging

Wireless sensor networks are also used for the collection of data for monitoring of environmental information, this can be as simple as the monitoring of the temperature in a fridge to the level of water in overflow tanks in nuclear power plants. The statistical information can then be used to show how systems have been working. The advantage of WSNs over conventional loggers is the "live" data feed that is possible.

Water/Waste Water Monitoring

Monitoring the quality and level of water includes many activities such as checking the quality of underground or surface water and ensuring a country's water infrastructure for the benefit of both human and animal. It may be used to protect the wastage of water,

Structural Health Monitoring

Wireless sensor networks can be used to monitor the condition of civil infrastructure and related geo-physical processes close to real time, and over long periods through data logging, using appropriately interfaced sensors.

Wine Production

Wireless sensor networks are used to monitor wine production, both in the field and the cellar.

Characteristics

The main characteristics of a WSN include:

- Power consumption constraints for nodes using batteries or energy harvesting

- Ability to cope with node failures (resilience)

- Some mobility of nodes

- Heterogeneity of nodes

- Scalability to large scale of deployment

- Ability to withstand harsh environmental conditions

- Ease of use

- Cross-layer design

Cross-layer is becoming an important studying area for wireless communications. In addition, the traditional layered approach presents three main problems:

1. Traditional layered approach cannot share different information among different layers, which leads to each layer not having complete information. The traditional layered approach cannot guarantee the optimization of the entire network.

2. The traditional layered approach does not have the ability to adapt to the environmental change.

3. Because of the interference between the different users, access conflicts, fading, and the change of environment in the wireless sensor networks, traditional layered approach for wired networks is not applicable to wireless networks.

So the cross-layer can be used to make the optimal modulation to improve the transmission performance, such as data rate, energy efficiency, QoS (Quality of Service), etc.. Sensor nodes can be imagined as small computers which are extremely basic in terms of their interfaces and their components. They usually consist of a *processing unit* with limited computational power and limited memory, *sensors* or MEMS (including specific conditioning circuitry), a *communication device* (usually radio transceivers or alternatively optical), and a power source usually in the form of a battery. Other possible inclusions are energy harvesting modules, secondary ASICs, and possibly secondary communication interface (e.g. RS-232 or USB).

The base stations are one or more components of the WSN with much more computational, energy and communication resources. They act as a gateway between sensor nodes and the end user as they typically forward data from the WSN on to a server. Other special components in routing based networks are routers, designed to compute, calculate and distribute the routing tables.

Platforms

Hardware

One major challenge in a WSN is to produce *low cost* and *tiny* sensor nodes. There are an increasing number of small companies producing WSN hardware and the commercial situation can be compared to home computing in the 1970s. Many of the nodes are still in the research and development stage, particularly their software. Also inherent to sensor network adoption is the use of very low power methods for radio communication and data acquisition.

In many applications, a WSN communicates with a Local Area Network or Wide Area Network through a gateway. The Gateway acts as a bridge between the WSN and the other network. This enables data to be stored and processed by devices with more resources, for example, in a remotely located server.

Software

Energy is the scarcest resource of WSN nodes, and it determines the lifetime of WSNs. WSNs may be deployed in large numbers in various environments, including remote and hostile regions, where ad hoc communications are a key component. For this reason, algorithms and protocols need to address the following issues:

- Increased lifespan

- Robustness and fault tolerance

- Self-configuration

Lifetime maximization: Energy/Power Consumption of the sensing device should be minimized and sensor nodes should be energy efficient since their limited energy resource determines their lifetime. To conserve power, wireless sensor nodes normally power off both the radio transmitter and the radio receiver when not in use.

Operating Systems

Operating systems for wireless sensor network nodes are typically less complex than general-purpose operating systems. They more strongly resemble embedded systems, for two reasons. First, wireless sensor networks are typically deployed with a particular application in mind, rather than as a general platform. Second, a need for low costs and low power leads most wireless sensor nodes to have low-power microcontrollers ensuring that mechanisms such as virtual memory are either unnecessary or too expensive to implement.

It is therefore possible to use embedded operating systems such as eCos or uC/OS for sensor networks. However, such operating systems are often designed with real-time properties.

TinyOS is perhaps the first operating system specifically designed for wireless sensor networks. TinyOS is based on an event-driven programming model instead of multithreading. TinyOS programs are composed of *event handlers* and *tasks* with run-to-completion semantics. When an external event occurs, such as an incoming data packet or a sensor reading, TinyOS signals the appropriate event handler to handle the event. Event handlers can post tasks that are scheduled by the TinyOS kernel some time later.

LiteOS is a newly developed OS for wireless sensor networks, which provides UNIX-like abstraction and support for the C programming language.

Contiki is an OS which uses a simpler programming style in C while providing advances such as 6LoWPAN and Protothreads.

Online Collaborative Sensor Data Management Platforms

Online collaborative sensor data management platforms are on-line database services that allow sensor owners to register and connect their devices to feed data into an online database for storage and also allow developers to connect to the database and build their own applications based on that data. Examples include Xively and the Wikisensing platform. Such platforms simplify online collaboration between users over diverse data sets ranging from energy and environment data to that collected from transport services. Other services include allowing developers to embed real-time graphs & widgets in websites; analyse and process historical data pulled from the data feeds; send real-time alerts from any datastream to control scripts, devices and environments.

The architecture of the Wikisensing system describes the key components of such systems to include APIs and interfaces for online collaborators, a middleware containing the business logic needed for the sensor data management and processing and a storage model suitable for the efficient storage and retrieval of large volumes of data. 0525162632

Simulation of Wsns

At present, agent-based modeling and simulation is the only paradigm which allows the simulation of complex behavior in the environments of wireless sensors (such as flocking). Agent-based simulation of wireless sensor and ad hoc networks is a relatively new paradigm. Agent-based modelling was originally based on social simulation.

Network simulators like OPNET, NetSim and NS2 can be used to simulate a wireless sensor network.

Other Concepts

Distributed Sensor Network

If a centralised architecture is used in a sensor network and the central node fails, then the entire network will collapse, however the reliability of the sensor network can be increased by using a distributed control architecture. Distributed control is used in WSNs for the following reasons:

1. Sensor nodes are prone to failure,

2. For better collection of data,

3. To provide nodes with backup in case of failure of the central node.

There is also no centralised body to allocate the resources and they have to be self organized.

Data Integration and Sensor Web

The data gathered from wireless sensor networks is usually saved in the form of numerical data in a central base station. Additionally, the Open Geospatial Consortium (OGC) is specifying standards for interoperability interfaces and metadata encodings that enable real time integration of heterogeneous sensor webs into the Internet, allowing any individual to monitor or control wireless sensor networks through a web browser.

In-network processing

To reduce communication costs some algorithms remove or reduce nodes' redundant sensor information and avoid forwarding data that is of no use. As nodes can inspect the data they forward, they can measure averages or directionality for example of readings from other nodes. For example, in sensing and monitoring applications, it is generally the case that neighboring sensor nodes monitoring an environmental feature typically register similar values. This kind of data redundancy due to the spatial correlation between sensor observations inspires techniques for in-network data aggregation and mining

Sonar

French F70 type frigates (here, *La Motte-Picquet*) are fitted with VDS (Variable Depth Sonar) type DUBV43 or DUBV43C towed sonars

Sonar (originally an acronym for SOund Navigation And Ranging) is a technique that uses sound propagation (usually underwater, as in submarine navigation) to navigate, communicate with or detect objects on or under the surface of the water, such as other vessels. Two types of technology share the name "sonar": *passive* sonar is essentially listening for the sound made by vessels; *active* sonar is emitting pulses of sounds and listening for echoes. Sonar may be used as a means of acoustic location and of measurement of the echo characteristics of "targets" in the water. Acoustic location in air was used before the introduction of radar. Sonar may also be used in air for robot navigation, and SODAR (an upward looking in-air sonar) is used for atmospheric investigations. The term *sonar* is also used for the equipment used to generate and receive the sound. The acoustic frequencies used in sonar systems vary from very low (infrasonic) to extremely high (ultrasonic). The study of underwater sound is known as underwater acoustics or hydroacoustics.

Sonar image of shipwreck of the Latvian Naval Forces ship *Virsaitis* in Estonian waters.

History

Although some animals (dolphins and bats) have used sound for communication and object detection for millions of years, use by humans in the water is initially recorded by Leonardo da Vinci in 1490: a tube inserted into the water was said to be used to detect vessels by placing an ear to the tube.

In the 19th century an underwater bell was used as an ancillary to lighthouses to provide warning of hazards.

The use of sound to 'echo locate' underwater in the same way as bats use sound for aerial navigation seems to have been prompted by the *Titanic* disaster of 1912. The world's first patent for an underwater echo ranging device was filed at the British Patent Office by English meteorologist Lewis Richardson a month after the sinking of the Titanic, and a German physicist Alexander Behm obtained a patent for an echo sounder in 1913.

The Canadian engineer Reginald Fessenden, while working for the Submarine Signal Company in Boston, built an experimental system beginning in 1912, a system later tested in Boston Harbor, and finally in 1914 from the U.S. Revenue (now Coast Guard) Cutter Miami on the Grand Banks off Newfoundland Canada. In that test, Fessenden demonstrated depth sounding, underwater communications (Morse code) and echo ranging (detecting an iceberg at two miles (3 km) range). The so-called Fessenden oscillator, at ca. 500 Hz frequency, was unable to determine the bearing of the berg due to the 3 metre wavelength and the small dimension of the transducer's radiating face (less than 1 metre in diameter). The ten Montreal-built British H class submarines launched in 1915 were equipped with a Fessenden oscillator.

During World War I the need to detect submarines prompted more research into the use of sound. The British made early use of underwater listening devices called hydrophones, while the French physicist Paul Langevin, working with a Russian immigrant electrical engineer, Constantin Chilowsky, worked on the development of active sound devices for detecting submarines in 1915. Although piezoelectric and magnetostrictive transducers later superseded the electrostatic transducers they used, this work influenced future designs. Lightweight sound-sensitive plastic film and fibre optics have been used for hydrophones (acousto-electric transducers for in-water use), while Terfenol-D and PMN (lead magnesium niobate) have been developed for projectors.

ASDIC

In 1916, under the British Board of Invention and Research, Canadian physicist Robert William Boyle took on the active sound detection project with A B Wood, producing a prototype for testing in mid-1917. This work, for the Anti-Submarine Division of the British Naval Staff, was undertaken in utmost secrecy, and used quartz piezoelectric crystals to produce the world's first practical underwater active sound detection apparatus. To maintain secrecy no mention of sound experimentation or quartz was made - the word used to describe the early work ('supersonics') was changed to 'ASD'ics, and the quartz material to 'ASD'ivite: "ASD" for "Anti-Submarine Division", hence the British acronym *ASDIC*. In 1939, in response to a question from the Oxford English Dictionary, the Admiralty made up the story that it stood for 'Allied Submarine Detection Investigation Committee', and this is still widely believed, though no committee bearing this name has been found in the Admiralty archives.

By 1918, both France and Britain had built prototype active systems. The British tested their ASDIC on HMS *Antrim* in 1920, and started production in 1922. The 6th Destroyer Flotilla had ASDIC-equipped vessels in 1923. An anti-submarine school, HMS *Osprey*, and a training flotilla of four vessels were established on Portland in 1924. The US Sonar QB set arrived in 1931.

ASDIC display unit ca. 1944

By the outbreak of World War II, the Royal Navy had five sets for different surface ship classes, and others for submarines, incorporated into a complete anti-submarine attack system. The effectiveness of early ASDIC was hamstrung by the use of the depth charge as an anti-submarine weapon. This required an attacking vessel to pass over a submerged contact before dropping charges over the stern, resulting in a loss of ASDIC contact in the moments leading up to attack. The hunter was effectively firing blind, during which time a submarine commander could take evasive action. This situation was remedied by using several ships cooperating and by the adoption of "ahead throwing weapons", such as Hedgehog and later Squid, which projected warheads at a target ahead of the attacker and thus still in ASDIC contact. Developments during the war resulted in British ASDIC sets which used several different shapes of beam, continuously covering blind spots. Later, acoustic torpedoes were used.

At the start of World War II, British ASDIC technology was transferred for free to the United States. Research on ASDIC and underwater sound was expanded in the UK and in the US. Many new types of military sound detection were developed. These included sonobuoys, first developed by the British in 1944 under the codename *High Tea*, dipping/dunking sonar and mine detection sonar. This work formed the basis for post war developments related to countering the nuclear submarine. Work on sonar had also been carried out in the Axis countries, notably in Germany, which included countermeasures. At the end of World War II this German work was assimilated by Britain and the US. Sonars have continued to be developed by many countries, including Russia, for both military and civil uses. In recent years the major military development has been the increasing interest in low frequency active systems.

SONAR

During the 1930s American engineers developed their own underwater sound detection technology and important discoveries were made, such as thermoclines, that would help future development. After technical information was exchanged between the two countries during the Second World War, Americans began to use the term *SONAR* for their systems, coined as the equivalent of RADAR.

Materials and Designs

There was little progress in development from 1915 to 1940. In 1940, the US sonars typically consisted of a magnetostrictive transducer and an array of nickel tubes connected to a 1-foot-diameter steel plate attached back to back to a Rochelle salt crystal in a spherical housing. This assembly penetrated the ship hull and was manually rotated to the desired angle. The piezoelectric Rochelle salt crystal had better parameters, but the magnetostrictive unit was much more reliable. Early WW2 losses prompted rapid research in the field, pursuing both improvements in magnetostrictive transducer parameters and Rochelle salt reliability. Ammonium dihydrogen phosphate (ADP), a superior alternative, was found as a replacement for Rochelle salt; the first application was a replacement of the 24 kHz Rochelle salt transducers. Within nine months, Rochelle salt was obsolete. The ADP manufacturing facility grew from few dozen personnel in early 1940 to several thousands in 1942.

One of the earliest application of ADP crystals were hydrophones for acoustic mines; the crystals were specified for low frequency cutoff at 5 Hz, withstanding mechanical shock for deployment from aircraft from 3,000 m (10,000 ft), and ability to survive neighbouring mine explosions. One of key features of ADP reliability is its zero aging characteristics; the crystal keeps its parameters even over prolonged storage.

Another application was for acoustic homing torpedoes. Two pairs of directional hydrophones were mounted on the torpedo nose, in the horizontal and vertical plane; the difference signals from the pairs were used to steer the torpedo left-right and up-down. A countermeasure was developed: the targeted submarine discharged an effervescent chemical, and the torpedo went after the noisier fizzy decoy. The counter-countermeasure was a torpedo with active sonar – a transducer was added to the torpedo nose, and the microphones were listening for its reflected periodic tone bursts. The transducers comprised identical rectangular crystal plates arranged to diamond-shaped areas in staggered rows.

Passive sonar arrays for submarines were developed from ADP crystals. Several crystal assemblies were arranged in a steel tube, vacuum-filled with castor oil, and sealed. The tubes then were mounted in parallel arrays.

The standard US Navy scanning sonar at the end of the World War II operated at 18 kHz, using an array of ADP crystals. Desired longer range however required use of lower frequencies. The required dimensions were too big for ADP crystals, so in the early 1950s magnetostrictive and barium titanate piezoelectric systems were developed, but these had problems achieving uniform impedance characteristics and the beam pattern suffered. Barium titanate was then replaced with more stable lead zirconate titanate (PZT), and the frequency was lowered to 5 kHz. The US fleet used this material in the AN/SQS-23 sonar for several decades. The SQS-23 sonar first used magnetostrictive nickel transducers, but these weighed several tons and nickel was expensive and considered a critical material; piezoelectric transducers were therefore substituted. The sonar was a large array of 432 individual transducers. At first the transducers were unreliable, showing mechanical and electrical failures and deteriorating soon after installation; they were also produced by several vendors, had different designs, and their characteristics were different enough to impair the array's performance. The policy to allow repair of individual transducers was then sacrificed, and "expendable modular design", sealed non-repairable modules, was chosen instead, eliminating the problem with seals and other extraneous mechanical parts.

The Imperial Japanese Navy at the onset of WW2 used projectors based on quartz. These were big and heavy, especially if designed for lower frequencies; the one for Type 91 set, operating at 9 kHz, had a diameter of 30 inches and was driven by an oscillator with 5 kW power and 7 kV of output amplitude. The Type 93 projectors consisted of solid sandwiches of quartz, assembled into spherical cast iron bodies. The Type 93 sonars were later replaced with Type 3, which followed German design and used magnetostrictive projectors; the projectors consisted of two rectangular identical independent units in a cast iron rectangular body about 16×9 inches. The exposed area was half the wavelength wide and three wavelengths high. The magnetostrictive cores were made from 4 mm stampings of nickel, and later of an iron-aluminium alloy with aluminium content between 12.7 and 12.9%. The power was provided from a 2 kW at 3.8 kV, with polarization from a 20 V/8 A DC source.

The passive hydrophones of the Imperial Japanese Navy were based on moving coil design, Rochelle salt piezo transducers, and carbon microphones.

Magnetostrictive transducers were pursued after WW2 as an alternative to piezoelectric ones. Nickel scroll-wound ring transducers were used for high-power low-frequency operations, with size up to 13 feet in diameter, probably the largest individual sonar transducers ever. The advantage of metals is their high tensile strength and low input electrical impedance, but they have electrical losses and lower coupling coefficient than PZT, whose tensile strength can be increased by prestressing. Other materials were also tried; nonmetallic ferrites were promising for their low electrical conductivity resulting in low eddy current losses, Metglas offered high coupling coefficient, but they were inferior to PZT overall. In the 1970s, compounds of rare earths and iron were discovered with superior magnetomechanic properties, namely the Terfenol-D alloy. This made possible new designs, e.g. a hybrid magnetostrictive-piezoelectric transducer. The most recent sch material is Galfenol.

Other types of transducers include variable reluctance (or moving armature, or electromagnetic) transducers, where magnetic force acts on the surfaces of gaps, and moving coil (or electrodynamic) transducers, similar to conventional speakers; the latter are used in underwater sound calibration, due to their very low resonance frequencies and flat broadband characteristics above them.

Active Sonar

Active sonar uses a sound transmitter and a receiver. When the two are in the same place it is monostatic operation. When the transmitter and receiver are separated it is bistatic operation. When more transmitters (or more receivers) are used, again spatially separated, it is multistatic operation. Most sonars are used monostatically with the same array often being used for transmission and reception. Active sonobuoy fields may be operated multistatically.

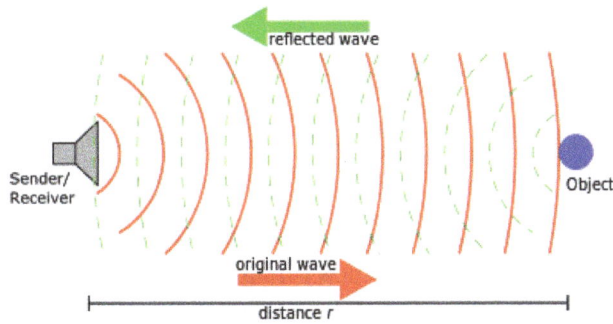

Principle of an active sonar

Active sonar creates a pulse of sound, often called a "ping", and then listens for reflections (echo) of the pulse. This pulse of sound is generally created electronically using a sonar projector consisting of a signal generator, power amplifier and electro-acoustic transducer/array. A beamformer is usually employed to concentrate the acoustic power into a beam, which may be swept to cover the required search angles. Generally, the electro-acoustic transducers are of the Tonpilz type and their design may be optimised to achieve maximum efficiency over the widest bandwidth, in order to optimise performance of the overall system. Occasionally, the acoustic pulse may be created by other means, e.g. (1) chemically using explosives, or (2) airguns or (3) plasma sound sources.

To measure the distance to an object, the time from transmission of a pulse to reception is measured and converted into a range by knowing the speed of sound. To measure the bearing, several hydrophones are used, and the set measures the relative arrival time to each, or with an array of hydrophones, by measuring the relative amplitude in beams formed through a process called beamforming. Use of an array reduces the spatial response so that to provide wide cover multi-beam systems are used. The target signal (if present) together with noise is then passed through various forms of signal processing, which for simple sonars may be just energy measurement. It is then presented to some form of decision device that calls the output either the required signal or noise. This decision device may be an operator with headphones or a display, or in more sophisticated sonars this function may be carried out by software. Further processes may be carried out to classify the target and localise it, as well as measuring its velocity.

The pulse may be at constant frequency or a chirp of changing frequency (to allow pulse compression on reception). Simple sonars generally use the former with a filter wide enough to cover possible Doppler changes due to target movement, while more complex ones generally include the latter technique. Since digital processing became available pulse compression has usually been implemented using digital correlation techniques. Military sonars often have multiple beams to provide all-round cover while simple ones only cover a narrow arc, although the beam may be rotated, relatively slowly, by mechanical scanning.

Particularly when single frequency transmissions are used, the Doppler effect can be used to measure the radial speed of a target. The difference in frequency between the transmitted and received signal is measured and converted into a velocity. Since Doppler shifts can be introduced by either receiver or target motion, allowance has to be made for the radial speed of the searching platform.

One useful small sonar is similar in appearance to a waterproof flashlight. The head is pointed into the water, a button is pressed, and the device displays the distance to the target. Another variant is a "fishfinder" that shows a small display with shoals of fish. Some civilian sonars (which are not designed for stealth) approach active military sonars in capability, with quite exotic three-dimensional displays of the area near the boat.

When active sonar is used to measure the distance from the transducer to the bottom, it is known as echo sounding. Similar methods may be used looking upward for wave measurement.

Active sonar is also used to measure distance through water between two sonar transducers or a combination of a hydrophone (underwater acoustic microphone) and projector (underwater acoustic speaker). A transducer is a device that can transmit and receive acoustic signals ("pings"). When a hydrophone/transducer receives a specific interrogation signal it responds by transmitting a specific reply signal. To measure distance, one transducer/projector transmits an interrogation signal and measures the time between this transmission and the receipt of the other transducer/hydrophone reply. The time difference, scaled by the speed of sound through water and divided by two, is the distance between the two platforms. This technique, when used with multiple transducers/hydrophones/projectors, can calculate the relative positions of static and moving objects in water.

In combat situations, an active pulse can be detected by an opponent and will reveal a submarine's position.

A very directional, but low-efficiency, type of sonar (used by fisheries, military, and for port security) makes use of a complex nonlinear feature of water known as non-linear sonar, the virtual transducer being known as a *parametric array*.

Project Artemis

Project Artemis was a one-of-a-kind low-frequency sonar for surveillance that was deployed off Bermuda for several years in the early 1960s. The active portion was deployed from a World War II tanker, and the receiving array was a built into a fixed position on an offshore bank.

Transponder

This is an active sonar device that receives a stimulus and immediately (or with a delay) retransmits the received signal or a predetermined one.

Performance Prediction

A sonar target is small relative to the sphere, centred around the emitter, on which it is located. Therefore, the power of the reflected signal is very low, several orders of magnitude less than the original signal. Even if the reflected signal was of the same power, the following example (using hypothetical values) shows the problem: Suppose a sonar system is capable of emitting a 10,000 W/m^2 signal at 1 m, and detecting a 0.001 W/m^2 signal. At 100 m the signal will be 1 W/m^2 (due to the inverse-square law). If the entire signal is reflected from a 10 m^2 target, it will be at 0.001 W/m^2 when it reaches the emitter, i.e. just detectable. However, the original signal will remain above 0.001 W/m^2 until 300 m. Any 10 m^2 target between 100 and 300 m using a similar or better system would be able to detect the pulse but would not be detected by the emitter. The detectors must be very sensitive to pick up the echoes. Since the original signal is much more powerful, it can be detected many times further than twice the range of the sonar (as in the example).

In active sonar there are two performance limitations, due to noise and reverberation. In general one or other of these will dominate so that the two effects can be initially considered separately.

In noise limited conditions at initial detection:

$$SL - 2TL + TS - (NL - DI) = DT$$

where SL is the source level, TL is the transmission loss (or propagation loss), TS is the target strength, NL is the noise level, DI is the directivity index of the array (an approximation to the array gain) and DT is the detection threshold.

In reverberation limited conditions at initial detection (neglecting array gain):

$$SL - 2TL + TS = RL + DT$$

where RL is the reverberation level and the other factors are as before.

Hand-held Sonar for Use by A Diver

- The LIMIS (= Limpet Mine Imaging Sonar) is a hand-held or ROV-mounted imaging sonar for use by a diver. Its name is because it was designed for patrol divers (combat frogmen or Clearance Divers) to look for limpet mines in low visibility water.ities

- The LUIS (= Lensing Underwater Imaging System) is another imaging sonar for use by a diver.

- There is or was a small flashlight-shaped handheld sonar for divers, that merely displays range.

- For the INSS = Integrated Navigation Sonar System

Passive Sonar

Passive sonar listens without transmitting. It is often employed in military settings, although it is also used in science applications, *e.g.*, detecting fish for presence/absence studies in various aquatic environments. In the very broadest usage, this term can encompass virtually any analytical technique involving remotely generated sound, though it is usually restricted to techniques applied in an aquatic environment.

Identifying Sound Sources

Passive sonar has a wide variety of techniques for identifying the source of a detected sound. For example, U.S. vessels usually operate 60 Hz alternating current power systems. If transformers or generators are mounted without proper vibration insulation from the hull or become flooded, the 60 Hz sound from the windings can be emitted from the submarine or ship. This can help to identify its nationality, as all European submarines and nearly every other nation's submarine have 50 Hz power systems. Intermittent sound sources (such as a wrench being dropped) may also be detectable to passive sonar. Until fairly recently, an experienced, trained operator identified signals, but now computers may do this.

Passive sonar systems may have large sonic databases, but the sonar operator usually finally classifies the signals manually. A computer system frequently uses these databases to identify classes of ships, actions (i.e. the speed of a ship, or the type of weapon released), and even particular ships. Publications for classification of sounds are provided by and continually updated by the US Office of Naval Intelligence.

Noise Limitations

Passive sonar on vehicles is usually severely limited because of noise generated by the vehicle. For this reason, many submarines operate nuclear reactors that can be cooled without pumps, using silent convection, or fuel cells or batteries, which can also run silently. Vehicles' propellers are also designed and precisely machined to emit minimal noise. High-speed propellers often create tiny bubbles in the water, and this cavitation has a distinct sound.

The sonar hydrophones may be towed behind the ship or submarine in order to reduce the effect of noise generated by the watercraft itself. Towed units also combat the thermocline, as the unit may be towed above or below the thermocline.

The display of most passive sonars used to be a two-dimensional waterfall display. The horizontal direction of the display is bearing. The vertical is frequency, or sometimes time. Another display technique is to color-code frequency-time information for bearing. More recent displays are generated by the computers, and mimic radar-type plan position indicator displays.

Performance Prediction

Unlike active sonar, only one way propagation is involved. Because of the different signal processing used, the minimum detectable signal to noise ratio will be different. The equation for determining the performance of a passive sonar is:

$$SL - TL = NL - DI + DT$$

where SL is the source level, TL is the transmission loss, NL is the noise level, DI is the directivity index of the array (an approximation to the array gain) and DT is the detection threshold. The figure of merit of a passive sonar is:

$$FOM = SL + DI - (NL + DT).$$

Performance Factors

The detection, classification and localisation performance of a sonar depends on the environment and the receiving equipment, as well as the transmitting equipment in an active sonar or the target radiated noise in a passive sonar.

Sound Propagation

Sonar operation is affected by variations in sound speed, particularly in the vertical plane. Sound travels more slowly in fresh water than in sea water, though the difference is small. The speed is determined by the water's bulk modulus and mass density. The bulk modulus is affected by temperature, dissolved impurities (usually salinity), and pressure. The density effect is small. The speed of sound (in feet per second) is approximately:

4388 + (11.25 × temperature (in °F)) + (0.0182 × depth (in feet)) + salinity (in parts-per-thousand).

This empirically derived approximation equation is reasonably accurate for normal temperatures, concentrations of salinity and the range of most ocean depths. Ocean temperature varies with depth, but at between 30 and 100 meters there is often a marked change, called the thermocline, dividing the warmer surface water from the cold, still waters that make up the rest of the ocean. This can frustrate sonar, because a sound originating on one side of the thermocline tends to be bent, or refracted, through the thermocline. The thermocline may be present in shallower coastal waters. However, wave action will often mix the water column and eliminate the thermocline. Water pressure also affects sound propagation: higher pressure increases the sound speed, which causes the sound waves to refract away from the area of higher sound speed. The mathematical model of refraction is called Snell's law.

If the sound source is deep and the conditions are right, propagation may occur in the 'deep sound channel'. This provides extremely low propagation loss to a receiver in the channel. This is because of sound trapping in the channel with no losses at the boundaries. Similar propagation can occur in the 'surface duct' under suitable conditions. However, in this case there are reflection losses at the surface.

In shallow water propagation is generally by repeated reflection at the surface and bottom, where considerable losses can occur.

Sound propagation is affected by absorption in the water itself as well as at the surface and bottom. This absorption depends upon frequency, with several different mechanisms in sea water. Long-range sonar uses low frequencies to minimise absorption effects.

The sea contains many sources of noise that interfere with the desired target echo or signature. The main noise sources are waves and shipping. The motion of the receiver through the water can also cause speed-dependent low frequency noise.

Scattering

When active sonar is used, scattering occurs from small objects in the sea as well as from the bottom and surface. This can be a major source of interference. This acoustic scattering is analogous to the scattering of the light from a car's headlights in fog: a high-intensity pencil beam will penetrate the fog to some extent, but broader-beam headlights emit much light in unwanted directions, much of which is scattered back to the observer, overwhelming that reflected from the target ("white-out"). For analogous reasons active sonar needs to transmit in a narrow beam to minimise scattering.

Target Characteristics

The sound *reflection* characteristics of the target of an active sonar, such as a submarine, are known as its target strength. A complication is that echoes are also obtained from other objects in the sea such as whales, wakes, schools of fish and rocks.

Passive sonar detects the target's *radiated* noise characteristics. The radiated spectrum comprises a continuous spectrum of noise with peaks at certain frequencies which can be used for classification.

Countermeasures

Active (powered) countermeasures may be launched by a submarine under attack to raise the noise level, provide a large false target, and obscure the signature of the submarine itself.

Passive (i.e., non-powered) countermeasures include:

- Mounting noise-generating devices on isolating devices.

- Sound-absorbent coatings on the hulls of submarines, for example anechoic tiles.

Military Applications

Modern naval warfare makes extensive use of both passive and active sonar from water-borne vessels, aircraft and fixed installations. Although active sonar was used by surface craft in World War II, submarines avoided the use of active sonar due to the potential for revealing their presence and position to enemy forces. However, the advent of modern signal-processing enabled the use of passive sonar as a primary means for search and detection operations. In 1987 a division of Japanese company Toshiba reportedly sold machinery to the Soviet Union that allowed their submarine propeller blades to be milled so that they became radically quieter, making the newer generation of submarines more difficult to detect.

The use of active sonar by a submarine to determine bearing is extremely rare and will not necessarily give high quality bearing or range information to the submarines fire control team. However, use of active sonar on surface ships is very common and is used by submarines when the tactical situation dictates it is more important to determine the position of a hostile submarine than conceal their own position. With surface ships, it might be assumed that the threat is already tracking the ship with satellite data as any vessel around the emitting sonar will detect the emission. Having heard the signal, it is easy to identify the sonar equipment used (usually with its frequency) and its position (with the sound wave's energy). Active sonar is similar to radar in that, while it allows detection of targets at a certain range, it also enables the emitter to be detected at a far greater range, which is undesirable.

Since active sonar reveals the presence and position of the operator, and does not allow exact classification of targets, it is used by fast (planes, helicopters) and by noisy platforms (most surface ships) but rarely by submarines. When active sonar is used by surface ships or submarines, it is typically activated very briefly at intermittent periods to minimize the risk of detection. Consequently, active sonar is normally considered a backup to passive sonar. In aircraft, active sonar is used in the form of disposable sonobuoys that are dropped in the aircraft's patrol area or in the vicinity of possible enemy sonar contacts.

Passive sonar has several advantages, most importantly that it is silent. If the target radiated noise level is high enough, it can have a greater range than active sonar, and allows the target to be identified. Since any motorized object makes some noise, it may in principle be detected, depending on the level of noise emitted and the ambient noise level in the area, as well as the technology used. To simplify, passive sonar "sees" around the ship using it. On a submarine, nose-mounted passive sonar detects in directions of about 270°, centered on the ship's alignment, the hull-mounted array of about 160° on each side, and the towed array of a full 360°. The invisible areas are due to the ship's own interference. Once a signal is detected in a certain direction (which means that something makes sound in that direction, this is called broadband detection) it is possible to zoom in and analyze the signal received (narrowband analysis). This is generally done using a Fourier transform to show the different frequencies making up the sound. Since every engine makes a specific sound, it is straightforward to identify the object. Databases of unique engine sounds are part of what is known as *acoustic intelligence* or ACINT.

Another use of passive sonar is to determine the target's trajectory. This process is called Target Motion Analysis (TMA), and the resultant "solution" is the target's range, course, and speed. TMA is done by marking from which direction the sound comes at different times, and comparing the motion with that of the operator's own ship. Changes in relative motion are analyzed using standard geometrical techniques along with some assumptions about limiting cases.

Passive sonar is stealthy and very useful. However, it requires high-tech electronic components and is costly. It is generally deployed on expensive ships in the form of arrays to enhance detection. Surface ships use it to good effect; it is even better used by submarines, and it is also used by airplanes and helicopters, mostly to a "surprise effect", since submarines can hide under thermal layers. If a submarine's commander believes he is alone, he may bring his boat closer to the surface and be easier to detect, or go deeper and faster, and thus make more sound.

Examples of sonar applications in military use are given below. Many of the civil uses given in the following section may also be applicable to naval use.

Anti-submarine Warfare

Variable Depth Sonar and its winch

Until recently, ship sonars were usually with hull mounted arrays, either amidships or at the bow. It was soon found after their initial use that a means of reducing flow noise was required. The first were made of canvas on a framework, then steel ones were used. Now domes are usually made of reinforced plastic or pressurized rubber. Such sonars are primarily active in operation. An example of a conventional hull mounted sonar is the SQS-56.

Because of the problems of ship noise, towed sonars are also used. These also have the advantage of being able to be placed deeper in the water. However, there are limitations on their use in shallow water. These are called towed arrays (linear) or variable depth sonars (VDS) with 2/3D arrays. A problem is that the winches required to deploy/recover these are large and expensive. VDS sets are primarily active in operation while towed arrays are passive.

An example of a modern active/passive ship towed sonar is Sonar 2087 made by Thales Underwater Systems.

Torpedoes

Modern torpedoes are generally fitted with an active/passive sonar. This may be used to home directly on the target, but wake following torpedoes are also used. An early example of an acoustic homer was the Mark 37 torpedo.

Torpedo countermeasures can be towed or free. An early example was the German *Sieglinde* device while the *Bold* was a chemical device. A widely used US device was the towed AN/SLQ-25 Nixie while Mobile submarine simulator (MOSS) was a free device. A modern alternative to the Nixie system is the UK Royal Navy S2170 Surface Ship Torpedo Defence system.

Mines

Mines may be fitted with a sonar to detect, localize and recognize the required target. Further information is given in acoustic mine and an example is the CAPTOR mine.

Mine Countermeasures

Mine Countermeasure (MCM) Sonar, sometimes called "Mine and Obstacle Avoidance Sonar (MOAS)", is a specialized type of sonar used for detecting small objects. Most MCM sonars are hull mounted but a few types are VDS design. An example of a hull mounted MCM sonar is the Type 2193 while the SQQ-32 Mine-hunting sonar and Type 2093 systems are VDS designs.

Submarine Navigation

Submarines rely on sonar to a greater extent than surface ships as they cannot use radar at depth. The sonar arrays may be hull mounted or towed. Information fitted on typical fits is given in Oyashio class submarine and Swiftsure class submarine.

Aircraft

Helicopters can be used for antisubmarine warfare by deploying fields of active/passive sonobuoys or can operate dipping sonar, such as the AQS-13. Fixed wing aircraft can also deploy sonobuoys and have greater endurance and capacity to deploy them. Processing from the sonobuoys or Dipping Sonar can be on the aircraft or on ship. Dipping sonar has the advantage of being deployable to depths appropriate to daily conditions Helicopters have also been used for mine countermeasure missions using towed sonars such as the AQS-20A.

AN/AQS-13 Dipping sonar deployed from an H-3 Sea King.

Underwater Communications

Dedicated sonars can be fitted to ships and submarines for underwater communication.

Ocean Surveillance

For many years, the United States operated a large set of passive sonar arrays at various points in the world's oceans, collectively called Sound Surveillance System (SOSUS) and later Integrated Undersea Surveillance System (IUSS). A similar system is believed to have been operated by the Soviet Union. As permanently mounted arrays in the deep ocean were utilised, they were in very quiet conditions so long ranges could be achieved. Signal processing was carried out using powerful computers ashore. With the ending of the Cold War a SOSUS array has been turned over to scientific use.

In the United States Navy, a special badge known as the Integrated Undersea Surveillance System Badge is awarded to those who have been trained and qualified in its operation.

Underwater Security

Sonar can be used to detect frogmen and other scuba divers. This can be applicable around ships or at entrances to ports. Active sonar can also be used as a deterrent and/or disablement mechanism. One such device is the Cerberus system.

Hand-held Sonar

Limpet Mine Imaging Sonar (LIMIS) is a hand-held or ROV-mounted imaging sonar designed for patrol divers (combat frogmen or clearance divers) to look for limpet mines in low visibility water.

The LUIS is another imaging sonar for use by a diver.

Integrated Navigation Sonar System (INSS) is a small flashlight-shaped handheld sonar for divers that displays range.

Intercept Sonar

This is a sonar designed to detect and locate the transmissions from hostile active sonars. An example of this is the Type 2082 fitted on the British Vanguard class submarines.

Civilian Applications

Fisheries

Fishing is an important industry that is seeing growing demand, but world catch tonnage is falling as a result of serious resource problems. The industry faces a future of continuing worldwide consolidation until a point of sustainability can be reached. However, the consolidation of the fishing fleets are driving increased demands for sophisticated fish finding electronics such as sensors, sounders and sonars. Historically, fishermen have used many different techniques to find and harvest fish. However, acoustic technology has been one of the most important driving forces behind the development of the modern commercial fisheries.

Sound waves travel differently through fish than through water because a fish's air-filled swim bladder has a different density than seawater. This density difference allows the detection of schools of fish by using reflected sound. Acoustic technology is especially well suited for underwater applications since sound travels farther and faster underwater than in air. Today, commercial fishing vessels rely almost completely on acoustic sonar and sounders to detect fish. Fishermen also use active sonar and echo sounder technology to determine water depth, bottom contour, and bottom composition.

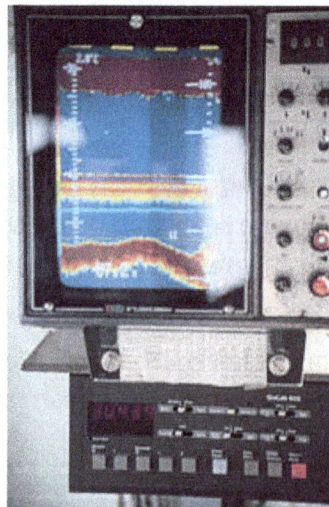

Cabin display of a fish finder sonar

Companies such as eSonar, Raymarine UK, Marport Canada, Wesmar, Furuno, Krupp, and Simrad make a variety of sonar and acoustic instruments for the deep sea commercial fishing industry. For example, net sensors take various underwater measurements and transmit the information back to a receiver on board a vessel. Each sensor is equipped with one or more acoustic transducers depending on its specific function. Data is transmitted from the sensors using wireless acoustic telemetry and is received by a hull mounted hydrophone. The analog signals are decoded and converted by a digital acoustic receiver into data which is transmitted to a bridge computer for graphical display on a high resolution monitor.

Echo Sounding

Echo sounding is a process used to determine the depth of water beneath ships and boats. A type of active sonar, echo sounding is the transmission of an acoustic pulse directly downwards to the seabed, measuring the time between transmission and echo return, after having hit the bottom and bouncing back to its ship of origin. The acoustic pulse is emitted by a transducer which receives the return echo as well. The depth measurement is calculated by multiplying the speed of sound in water(averaging 1,500 meters per second) by the time between emission and echo return.

The value of underwater acoustics to the fishing industry has led to the development of other acoustic instruments that operate in a similar fashion to echo-sounders but, because their function is slightly different from the initial model of the echo-sounder, have been given different terms.

Net Location

The net sounder is an echo sounder with a transducer mounted on the headline of the net rather than on the bottom of the vessel. Nevertheless, to accommodate the distance from the transducer to the display unit, which is much greater than in a normal echo-sounder, several refinements have to be made. Two main types are available. The first is the cable type in which the signals are sent along a cable. In this case there has to be the provision of a cable drum on which to haul, shoot and stow the cable during the different phases of the operation. The second type is the cable less net-sounder – such as Marport's Trawl Explorer - in which the signals are sent acoustically between the net and hull mounted receiver/hydrophone on the vessel. In this case no cable drum is required but sophisticated electronics are needed at the transducer and receiver.

The display on a net sounder shows the distance of the net from the bottom (or the surface), rather than the depth of water as with the echo-sounder's hull-mounted transducer. Fixed to the headline of the net, the footrope can usually be seen which gives an indication of the net performance. Any fish passing into the net can also be seen, allowing fine adjustments to be made to catch the most fish possible. In other fisheries, where the amount of fish in the net is important, catch sensor transducers are mounted at various positions on the cod-end of the net. As the cod-end fills up these catch sensor transducers are triggered one by one and this information is transmitted acoustically to display monitors on the bridge of the vessel. The skipper can then decide when to haul the net.

Modern versions of the net sounder, using multiple element transducers, function more like a sonar than an echo sounder and show slices of the area in front of the net and not merely the vertical view that the initial net sounders used.

The sonar is an echo-sounder with a directional capability that can show fish or other objects around the vessel.

ROV and UUV

Small sonars have been fitted to Remotely Operated Vehicles (ROV) and Unmanned Underwater Vehicles (UUV) to allow their operation in murky conditions. These sonars are used for looking ahead of the vehicle. The Long-Term Mine Reconnaissance System is an UUV for MCM purposes.

Vehicle Location

Sonars which act as beacons are fitted to aircraft to allow their location in the event of a crash in the sea. Short and Long Baseline sonars may be used for caring out the location, such as LBL.

Prosthesis for The Visually Impaired

In 2013 an inventor in the United States unveiled a "spider-sense" bodysuit, equipped with ultrasonic sensors and haptic feedback systems, which alerts the wearer of incoming threats; allowing them to respond to attackers even when blindfolded.

Scientific Applications

Biomass Estimation

Detection of fish, and other marine and aquatic life, and estimation their individual sizes or total biomass using active sonar techniques. As the sound pulse travels through water it encounters objects that are of different density or acoustic characteristics than the surrounding medium, such as fish, that reflect sound back toward the sound source. These echoes provide information on fish size, location, abundance and behavior. Data is usually processed and an-alysed using a variety of software such as Echoview.

Wave Measurement

An upward looking echo sounder mounted on the bottom or on a platform may be used to make measurements of wave height and period. From this statistics of the surface conditions at a location can be derived.

Water Velocity Measurement

Special short range sonars have been developed to allow measurements of water velocity.

Bottom Type Assessment

Sonars have been developed that can be used to characterise the sea bottom into, for example, mud, sand, and gravel. Relatively simple sonars such as echo sounders can be promoted to seafloor classification systems via add-on modules, converting echo parameters into sediment type. Different algorithms exist, but they are all based on changes in the energy or shape of the reflected sounder pings. Advanced substrate classification analysis can be achieved using calibrated (scientific) echosounders and parametric or fuzzy-logic analysis of the acoustic data.

Bathymetric Mapping

Side-scan sonars can be used to derive maps of seafloor topography (bathymetry) by moving the sonar across it just above the bottom. Low frequency sonars such as GLORIA have been used for continental shelf wide surveys while high frequency sonars are used for more detailed surveys of smaller areas.

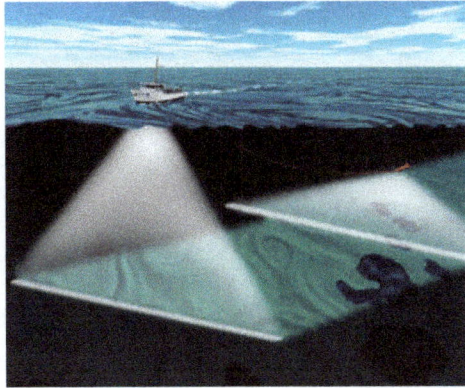

Graphic depicting hydrographic survey ship conducting multibeam and side-scan sonar operations

Sub-bottom Profiling

Powerful low frequency echo-sounders have been developed for providing profiles of the upper layers of the ocean bottom.

Synthetic Aperture Sonar

Various synthetic aperture sonars have been built in the laboratory and some have entered use in mine-hunting and search systems. An explanation of their operation is given in synthetic aperture sonar.

Parametric Sonar

Parametric sources use the non-linearity of water to generate the difference frequency between two high frequencies. A virtual end-fire array is formed. Such a projector has advantages of broad bandwidth, narrow beamwidth, and when fully developed and carefully measured it has no obvious sidelobes. Its major disadvantage is very low efficiency of only a few percent. P.J. Westervelt's seminal 1963 JASA paper summarizes the trends involved.

Sonar in Extraterrestrial Contexts

Use of sonar has been proposed for determining the depth of hydrocarbon seas on Titan.

Effect of Sonar on Marine Life

Effect on Marine Mammals

Research has shown that use of active sonar can lead to mass strandings of marine mammals. Beaked whales, the most common casualty of the strandings, have been shown to be highly sensitive to mid-frequency active sonar. Other marine mammals such as the blue whale also flee away from the source of the sonar, while naval activity was suggested to be the most probable cause of a mass stranding of dolphins. The US Navy, which part-funded some of studies, said the findings only showed behavioural responses to sonar, not actual harm, but "will evaluate the effectiveness of [their] marine mammal protective measures in light of new research findings."

A Humpback whale

Some marine animals, such as whales and dolphins, use echolocation systems, sometimes called *biosonar* to locate predators and prey. It is conjectured that active sonar transmitters could confuse these animals and interfere with basic biological functions such as feeding and mating.

Effect on Fish

High intensity sonar sounds can create a small temporary shift in the hearing threshold of some fish.

Frequencies and Resolutions

The frequencies of sonars range from infrasonic to above a megahertz. Generally, the lower frequencies have longer range, while the higher frequencies offer better resolution, and smaller size for a given directionality.

To achieve reasonable directionality, frequencies below 1 kHz generally require large size, usually achieved as towed arrays.

Low frequency sonars are loosely defined as 1–5 kHz, albeit some navies regard 5–7 kHz also as low frequency. Medium frequency is defined as 5–15 kHz. Another style of division considers low frequency to be under 1 kHz, and medium frequency at between 1–10 kHz.

American World War II era sonars operated at a relatively high frequency of 20–30 kHz, to achieve directionality with reasonably small transducers, with typical maximum operational range of 2500 yd. Postwar sonars used lower frequencies to achieve longer range; e.g. SQS-4 operated at 10 kHz with range up to 5000 yd. SQS-26 and SQS-53 operated at 3 kHz with range up to 20,000 yd; their domes had size of approx. a 60-ft personnel boat, an upper size limit for conventional hull sonars. Achieving larger sizes by conformal sonar array spread over the hull has not been effective so far, for lower frequencies linear or towed arrays are therefore used.

Japanese WW2 sonars operated at a range of frequencies. The Type 91, with 30 inch quartz projector, worked at 9 kHz. The Type 93, with smaller quartz projectors, operated at 17.5 kHz (model

5 at 16 or 19 kHz magnetostrictive) at powers between 1.7 and 2.5 kilowatts, with range of up to 6 km. The later Type 3, with German-design magnetostrictive transducers, operated at 13, 14.5, 16, or 20 kHz (by model), using twin transducers (except model 1 which had three single ones), at 0.2 to 2.5 kilowatts. The Simple type used 14.5 kHz magnetostrictive transducers at 0.25 kW, driven by capacitive discharge instead of oscillators, with range up to 2.5 km.

The sonar's resolution is angular; objects further apart will be imaged with lower resolutions than nearby ones.

Another source lists ranges and resolutions vs frequencies for sidescan sonars. 30 kHz provides low resolution with range of 1000–6000 m, 100 kHz gives medium resolution at 500–1000 m, 300 kHz gives high resolution at 150–500 m, and 600 kHz gives high resolution at 75–150 m. Longer range sonars are more adversely affected by nonhomogenities of water. Some environments, typically shallow waters near the coasts, have complicated terrain with many features; higher frequencies become necessary there.

As a specific example, the Sonar 2094 Digital, a towed fish capable of reaching depth of 1000 or 2000 meters, performs side-scanning at 114 kHz (600m range at each side, 50 by 1 degree beamwidth) and 410 kHz (150m range, 40 by 0.3 degree beamwidth), with 3 kW pulse power.

A JW Fishers system offers side-scanning at 1200 kHz with very high spatial resolution, optionally coupled with longer-range 600 kHz (range 200 ft at each side) or 100 kHz (up to 2000 ft per side, suitable for scanning large areas for big targets).

Echo Sounding

Illustration of echo sounding using a multibeam echosounder.

Echo sounding is a type of SONAR used to determine the depth of water by transmitting sound pulses into water. The time interval between emission and return of a pulse is recorded, which is used to determine the depth of water along with the speed of sound in water at the time. This information is then typically used for navigation purposes or in order to obtain depths for charting purposes. Echo sounding can also refer to hydroacoustic "echo sounders" defined as active sound in water (sonar) used to study fish. Hydroacoustic assessments have traditionally employed mobile surveys from boats to evaluate fish biomass and spatial distributions. Conversely, fixed-location techniques use stationary transducers to monitor passing fish.

The word *sounding* is used for all types of depth measurements, including those that don't use sound, and is unrelated in origin to the word *sound* in the sense of noise or tones. Echo sounding is a more rapid method of measuring depth than the previous technique of lowering a sounding line until it touched bottom.

Technique

Distance is measured by multiplying half the time from the signal's outgoing pulse to its return by the speed of sound in the water, which is approximately 1.5 kilometres per second [T÷2×(4700 feet per second or 1.5 kil per second)] For precise applications of echosounding, such as hydrography, the speed of sound must also be measured typically by deploying a sound velocity probe into the water. Echo sounding is effectively a special purpose application of sonar used to locate the bottom. Since a traditional pre-SI unit of water depth was the fathom, an instrument used for determining water depth is sometimes called a *fathometer*. The first practical fathometer was invented by Herbert Grove Dorsey and patented in 1928.

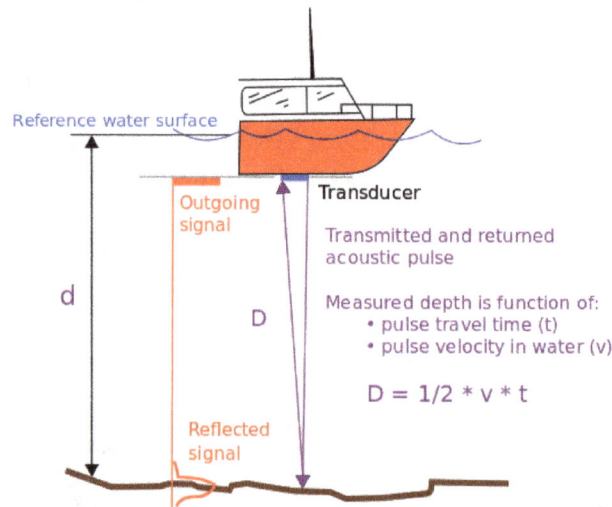

Figure 9-1. Acoustic depth measurement

Diagram showing the basic principle of echo sounding

Most charted ocean depths use an average or standard sound speed. Where greater accuracy is required average and even seasonal standards may be applied to ocean regions. For high accuracy depths, usually restricted to special purpose or scientific surveys, a sensor may be lowered to measure the temperature, pressure and salinity. These factors are used to calculate the actual sound speed in the local water column. This latter technique is regularly used by US Office of Coast Survey for navigational surveys of US coastal waters.

Common Use

As well as an aid to navigation (most larger vessels will have at least a simple depth sounder), echo sounding is commonly used for fishing. Variations in elevation often represent places where fish congregate. Schools of fish will also register. A fishfinder is an echo sounding device used by both recreational and commercial fishers.

Hydrography

In areas where detailed bathymetry is required, a precise echo sounder may be used for the work of hydrography. There are many considerations when evaluating such a system, not limited to the vertical accuracy, resolution, acoustic beamwidth of the transmit/receive beam and the acoustic frequency of the transducer.

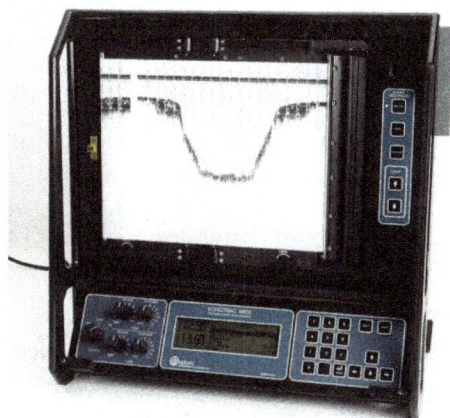

An example of a precision dual frequency echosounder, the Teledyne Odom MkIII

The majority of hydrographic echosounders are dual frequency, meaning that a low frequency pulse (typically around 24 kHz) can be transmitted at the same time as a high frequency pulse (typically around 200 kHz). As the two frequencies are discrete, the two return signals do not typically interfere with each other. There are many advantages of dual frequency echosounding, including the ability to identify a vegetation layer or a layer of soft mud on top of a layer of rock.

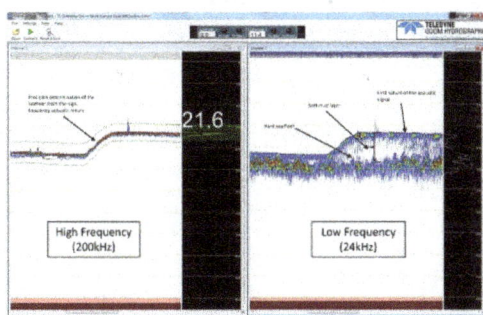

A screen grab of the difference between single and dual frequency echograms

Most hydrographic operations use a 200 kHz transducer, which is suitable for inshore work up to 100 metres in depth. Deeper water requires a lower frequency transducer as the acoustic signal of lower frequencies is less susceptible to attenuation in the water column. Commonly used frequencies for deep water sounding are 33 kHz and 24 kHz.

The beamwidth of the transducer is also a consideration for the hydrographer, as to obtain the best resolution of the data gathered a narrow beamwidth is preferable. This is especially important when sounding in deep water, as the resulting footprint of the acoustic pulse can be very large once it reaches a distant sea floor.

In addition to the single beam echo sounder, there are echo sounders that are capable of receiving many return "pings". These systems are detailed further in the section called multibeam echo-sounder.

Echo sounders are used in laboratory applications to monitor sediment transport, scour and erosion processes in scale models (hydraulic models, flumes etc.). These can also be used to create plots of 3D contours.

Standards for Hydrographic Echo Sounding

The required precision and accuracy of the hydrographic echo sounder is defined by the requirements of the International Hydrographic Organization (IHO) for surveys that are to be undertaken to IHO standards. These values are contained within IHO publication S44.

In order to meet these standards, the surveyor must consider not only the vertical and horizontal accuracy of the echo sounder and transducer, but the survey system as a whole. A motion sensor may be used, specifically the heave component (in single beam echosounding) to reduce soundings for the motion of the vessel experienced on the water's surface. Once all of the uncertainties of each sensor are established, the hydrographer will create an uncertainty budget to determine whether the survey system meets the requirements laid down by IHO.

Different hydrographic organisations will have their own set of field procedures and manuals to guide their surveyors to meet the required standards. Two examples are the US Army Corps of Engineers publication EM110-2-1003, and the NOAA 'Field Procedures Manual'.

History

German inventor Alexander Behm was granted German patent No. 282009 for the invention of echo sounding (*device for measuring depths of the sea and distances and headings of ships or obstacles by means of reflected sound waves*) on 22 July 1913.

Norwegian Inventor Hans Sundt Berggraf (1874-1941) published the same invention 8 years earlier, 8th of September 1904 in Teknisk Ukeblad.

Level Sensor

Level sensors detect the level of liquids and other fluids and fluidized solids, including slurries, granular materials, and powders that exhibit an upper free surface. Substances that flow become essentially horizontal in their containers (or other physical boundaries) because of gravity whereas most bulk solids pile at an angle of repose to a peak. The substance to be measured can be inside a container or can be in its natural form (e.g., a river or a lake). The level measurement can be either continuous or point values. Continuous level sensors measure level within a specified range and determine the exact amount of substance in a certain place, while point-level sensors only indicate whether the substance is above or below the sensing point. Generally the latter detect levels that are excessively high or low.

There are many physical and application variables that affect the selection of the optimal level monitoring method for industrial and commercial processes. The selection criteria include the physical: phase (liquid, solid or slurry), temperature, pressure or vacuum, chemistry, dielectric constant of medium, density (specific gravity) of medium, agitation (action), acoustical or electrical noise, vibration, mechanical shock, tank or bin size and shape. Also important are the application constraints: price, accuracy, appearance, response rate, ease of calibration or programming, physical size and mounting of the instrument, monitoring or control of continuous or discrete (point) levels. In short, level sensors are one of the very important sensors and play very important role in variety of consumer/ industrial applications. As with other type of sensors, level sensors are available or can be designed using variety of sensing principles. Selection of an appropriate type of sensor suiting to the application requirement is very important.

Point and Continuous Level Detection for Solids

A variety of sensors are available for point level detection of solids. These include vibrating, rotating paddle, mechanical (diaphragm), microwave (radar), capacitance, optical, pulsed-ultrasonic and ultrasonic level sensors.

Vibrating Point

These detect levels of very fine powders (bulk density: 0.02 g/cm^3 – 0.2 g/cm^3), fine powders (bulk density: 0.2 – 0.5 g/cm^3), and granular solids (bulk density: 0.5 g/cm^3 or greater). With proper selection of vibration frequency and suitable sensitivity adjustments, they can also sense the level of highly fluidized powders and electrostatic materials.

Single-probe vibrating level sensors are ideal for bulk powder level. Since only one sensing element contacts the powder, bridging between two probe elements is eliminated and media build-up is minimized. The vibration of the probe tends to eliminate build-up of material on the probe element. Vibrating level sensors are not affected by dust, static-charge build-up from dielectric powders, or changes in conductivity, temperature, pressure, humidity or moisture content. Tuning-fork style vibration sensors are another alternative. They tend to be less costly, but are prone to material buildup between the tines,

Rotating Paddle

Rotating paddle level sensors are a very old and established technique for bulk solid point level indication. The technique uses a low-speed gear motor that rotates a paddle wheel. When the paddle is stalled by solid materials, the motor is rotated on its shaft by its own torque until a flange mounted on the motor contacts a mechanical switch. The paddle can be constructed from a variety of materials, but tacky material must not be allowed to build up on the paddle. Build-up may occur if the process material becomes tacky because of high moisture levels or high ambient humidity in the hopper. For materials with very low weight per unit volume such as Pearlite, Bentonite or fly ash, special paddle designs and low-torque motors are used. Fine particles or dust must be prevented from penetrating the shaft bearings and motor by proper placement of the paddle in the hopper or bin and using appropriate seals.

Admittance-type

An RF Admittance level sensor uses a rod probe and RF source to measures the change in admittance. The probe is driven through a shielded coaxial cable to eliminate the effects of changing cable capacitance to ground. When the level changes around the probe, a corresponding change in the di-electric is observed. This changes the admittance of this imperfect capacitor and this change is measured to detect change of level.

Point Level Detection of Liquids

Magnetic and Mechanical Float

The principle behind magnetic, mechanical, cable, and other float level sensors often involves the opening or closing of a mechanical switch, either through direct contact with the switch, or magnetic operation of a reed. In other instances, such as magnetostrictive sensors, continuous monitoring is possible using a float principle.

With magnetically actuated float sensors, switching occurs when a permanent magnet sealed inside a float rises or falls to the actuation level. With a mechanically actuated float, switching occurs as a result of the movement of a float against a miniature (micro) switch. For both magnetic and mechanical float level sensors, chemical compatibility, temperature, specific gravity (density), buoyancy, and viscosity affect the selection of the stem and the float. For example, larger floats may be used with liquids with specific gravities as low as 0.5 while still maintaining buoyancy. The choice of float material is also influenced by temperature-induced changes in specific gravity and viscosity – changes that directly affect buoyancy.

Float-type sensors can be designed so that a shield protects the float itself from turbulence and wave motion. Float sensors operate well in a wide variety of liquids, including corrosives. When used for organic solvents, however, one will need to verify that these liquids are chemically compatible with the materials used to construct the sensor. Float-style sensors should not be used with high viscosity (thick) liquids, sludge or liquids that adhere to the stem or floats, or materials that contain contaminants such as metal chips; other sensing technologies are better suited for these applications.

A special application of float type sensors is the determination of interface level in oil-water separation systems. Two floats can be used with each float sized to match the specific gravity of the oil on one hand, and the water on the other. Another special application of a stem type float switch is the installation of temperature or pressure sensors to create a multi-parameter sensor. Magnetic float switches are popular for simplicity, dependability and low cost.

Pneumatic

Pneumatic level sensors are used where hazardous conditions exist, where there is no electric power or its use is restricted, and in applications involving heavy sludge or slurry. As the compression of a column of air against a diaphragm is used to actuate a switch, no process liquid contacts the sensor's moving parts. These sensors are suitable for use with highly viscous liquids such as grease, as well as water-based and corrosive liquids. This has the additional benefit of being a relatively low cost technique for point level monitoring.

Conductive

Conductive level sensors are ideal for the point level detection of a wide range of conductive liquids such as water, and is especially well suited for highly corrosive liquids such as caustic soda, hydrochloric acid, nitric acid, ferric chloride, and similar liquids. For those conductive liquids that are corrosive, the sensor's electrodes need to be constructed from titanium, Hastelloy B or C, or 316 stainless steel and insulated with spacers, separators or holders of ceramic, polyethylene and Teflon-based materials. Depending on their design, multiple electrodes of differing lengths can be used with one holder. Since corrosive liquids become more aggressive as temperature and pressure increase, these extreme conditions need to be considered when specifying these sensors.

Conductive level sensors use a low-voltage, current-limited power source applied across separate electrodes. The power supply is matched to the conductivity of the liquid, with higher voltage versions designed to operate in less conductive (higher resistance) mediums. The power source frequently incorporates some aspect of control, such as high-low or alternating pump control. A conductive liquid contacting both the longest probe (common) and a shorter probe (return) completes a conductive circuit. Conductive sensors are extremely safe because they use low voltages and currents. Since the current and voltage used is inherently small, for personal safety reasons, the technique is also capable of being made "Intrinsically Safe" to meet international standards for hazardous locations. Conductive probes have the additional benefit of being solid-state devices and are very simple to install and use. In some liquids and applications, maintenance can be an issue. The probe must continue to be conductive. If buildup insulates the probe from the medium, it will stop working properly. A simple inspection of the probe will require an ohmmeter connected across the suspect probe and the ground reference.

Typically, in most water and wastewater wells, the well itself with its ladders, pumps and other metal installations, provides a ground return. However, in chemical tanks, and other non-grounded wells, the installer must supply a ground return, typically an earth rod.

Sensors for Both Point Level Detection or Continuous Monitoring

Ultrasonic

Ultrasonic level sensors are used for non-contact level sensing of highly viscous liquids, as well as bulk solids. They are also widely used in water treatment applications for pump control and open channel flow measurement. The sensors emit high frequency (20 kHz to 200 kHz) acoustic waves that are reflected back to and detected by the emitting transducer.

Ultrasonic level sensors are also affected by the changing speed of sound due to moisture, temperature, and pressures. Correction factors can be applied to the level measurement to improve the accuracy of measurement.

Turbulence, foam, steam, chemical mists (vapors), and changes in the concentration of the process material also affect the ultrasonic sensor's response. Turbulence and foam prevent the sound wave from being properly reflected to the sensor; steam and chemical mists and vapors distort or absorb the sound wave; and variations in concentration cause changes in the amount of energy in the sound wave that is reflected back to the sensor. Stilling wells and wave guides are used to prevent errors caused by these factors.

Proper mounting of the transducer is required to ensure best response to reflected sound. In addition, the hopper, bin, or tank should be relatively free of obstacles such as weldments, brackets, or ladders to minimise false returns and the resulting erroneous response, although most modern systems have sufficiently "intelligent" echo processing to make engineering changes largely unnecessary except where an intrusion blocks the "line of sight" of the transducer to the target. Since the ultrasonic transducer is used both for transmitting and receiving the acoustic energy, it is subject to a period of mechanical vibration known as "ringing". This vibration must attenuate (stop) before the echoed signal can be processed. The net result is a distance from the face of the transducer that is blind and cannot detect an object. It is known as the "blanking zone", typically 150mm – 1m, depending on the range of the transducer.

The requirement for electronic signal processing circuitry can be used to make the ultrasonic sensor an intelligent device. Ultrasonic sensors can be designed to provide point level control, continuous monitoring or both. Due to the presence of a microprocessor and relatively low power consumption, there is also capability for serial communication from to other computing devices making this a good technique for adjusting calibration and filtering of the sensor signal, remote wireless monitoring or plant network communications. The ultrasonic sensor enjoys wide popularity due to the powerful mix of low price and high functionality.

Capacitance

Capacitance level sensors excel in sensing the presence of a wide variety of solids, aqueous and organic liquids, and slurries. The technique is frequently referred to as RF for the radio frequency signals applied to the capacitance circuit. The sensors can be designed to sense material with dielectric constants as low as 1.1 (coke and fly ash) and as high as 88 (water) or more. Sludges and slurries such as dehydrated cake and sewage slurry (dielectric constant approx. 50) and liquid chemicals such as quicklime (dielectric constant approx. 90) can also be sensed. Dual-probe capacitance level sensors can also be used to sense the interface between two immiscible liquids with substantially different dielectric constants, providing a solid state alternative to the aforementioned magnetic float switch for the "oil-water interface" application.

Since capacitance level sensors are electronic devices, phase modulation and the use of higher frequencies makes the sensor suitable for applications in which dielectric constants are similar. The sensor contains no moving parts, is rugged, simple to use, and easy to clean, and can be designed for high temperature and pressure applications. A danger exists from build-up and discharge of a high-voltage static charge that results from the rubbing and movement of low dielectric materials, but this danger can be eliminated with proper design and grounding.

Appropriate choice of probe materials reduces or eliminates problems caused by abrasion and corrosion. Point level sensing of adhesives and high-viscosity materials such as oil and grease can result in the build-up of material on the probe; however, this can be minimized by using a self-tuning sensor. For liquids prone to foaming and applications prone to splashing or turbulence, capacitance level sensors can be designed with splashguards or stilling wells, among other devices.

A significant limitation for capacitance probes is in tall bins used for storing bulk solids. The requirement for a conductive probe that extends to the bottom of the measured range is problematic. Long conductive cable probes (20 to 50 meters long), suspended into the bin or silo, are subject to tremendous mechanical tension due to the weight of the bulk powder in the silo and the friction applied to the cable. Such installations will frequently result in a cable breakage.

Optical Interface

Optical sensors are used for point level sensing of sediments, liquids with suspended solids, and liquid-liquid interfaces. These sensors sense the decrease or change in transmission of infrared light emitted from an infrared diode (LED). With the proper choice of construction materials and mounting location, these sensors can be used with aqueous, organic, and corrosive liquids.

A common application of economical infrared-based optical interface point level sensors is detecting the sludge/water interface in settling ponds. By using pulse modulation techniques and a high power infrared diode, one can eliminate interference from ambient light, operate the LED at a higher gain, and lessen the effects of build-up on the probe.

An alternate approach for continuous optical level sensing involves the use of a laser. Laser light is more concentrated and therefore is more capable of penetrating dusty or steamy environments. Laser light will reflect off most solid, liquid surfaces. The time of flight can be measured with precise timing circuitry, to determine the range or distance of the surface from the sensor. Lasers remain limited in use in industrial applications due to cost, and concern for maintenance. The optics must be frequently cleaned to maintain performance.

Microwave

Microwave sensors are ideal for use in moist, vaporous, and dusty environments as well as in applications in which temperatures and pressures vary. Microwaves (also frequently described as RADAR), will penetrate temperature and vapor layers that may cause problems for other techniques, such as ultrasonic. Microwaves are electromagnetic energy and therefore do not require air molecules to transmit the energy making them useful in vacuums. Microwaves, as electromagnetic energy, are reflected by objects with high conductive properties, like metal and conductive water. Alternately, they are absorbed in various degrees by 'low dielectric' or insulating mediums such as plastics, glass, paper, many powders and food stuffs and other solids.

Microwave sensors are executed in a wide variety of techniques. Two basic signal processing techniques are applied, each offering its own advantages: Pulsed or Time-Domain Reflectometry

(TDR) which is a measurement of time of flight divided by the speed of light, similar to ultrasonic level sensors, and Doppler systems employing FMCW techniques. Just as with ultrasonic level sensors, microwave sensors are executed at various frequencies, from 1 GHz to 60 GHz. Generally, the higher the frequency, the more accurate, and the more costly. Microwave is executed non-contact technique or guided. The first is done by monitoring a microwave signal that is transmitted through free space (including vacuum) and reflected back, or can be executed as a "radar on a wire" technique, generally known as Guided Wave Radar or Guided Microwave Radar. In the latter technique, performance generally improves in powders and low dielectric media that are not good reflectors of electromagnetic energy transmitted through a void (as in non-contact microwave sensors). But with the guided technique the same mechanical constraints exist that cause problems for the capacitance (RF) techniques mentioned previously by having a probe in the vessel.

Non contact microwave-based radar sensors are able to see through low conductivity 'microwave-transparent' (non-conductive) glass/plastic windows or vessel walls through which the microwave beam can be passed and measure a 'microwave reflective' (conductive) liquid inside (in the same way as to use a plastic bowl in a microwave oven). They are also largely unaffected by high temperature, pressure, vacuum or vibration. As these sensors do not require physical contact with the process material, so the transmitter /receiver can be mounted a safe distance above/from the process, even with an antenna extension of several metres to reduce temperature, yet still respond to the changes in level or distance changes e.g. they are ideal for measurement of molten metal products at over 1200 °C. Microwave transmitters also offer the same key advantage of ultrasonics: the presence of a microprocessor to process the signal, provide numerous monitoring, controls, communications, setup and diagnostic capabilities and are independent of changing density, viscosity and electrical properties. Additionally, they solve some of the application limitations of ultrasonics: operation in high pressure and vacuum, high temperatures, dust, temperature and vapor layers. Guided Wave Radars can measure in narrow confined spaces very successfully, as the guide element ensures correct transmission to and from the measured liquid. Applications such as inside stilling tubes or external bridles or cages, offer an excellent alternative to float or displacement devices, as they remove any moving parts or linkages and are unaffected by density changes or build up. They are also excellent with very low microwave reflectivity products like liquid gasses (LNG, LPG, Ammonia) which are stored at low temperatures/high pressures, although care needs to be taken on sealing arrangements and hazardous area approvals. On bulk solids and powders, GWR offers a great alternative to radar or ultrasonic sensors, but some care needs to be taken over cable wear and roof loading by the product movement.

One perceived major disadvantage of microwave or radar techniques for level monitoring is the relatively high price of such sensors and complex set up. However, price has reduced significantly over the last few years, to match those of longer range ultrasonics, with simplified set up of both techniques also improving ease of use.

Continuous Level Measurement of Liquids

Magnetostrictive

Magnetostrictive level sensors are similar to float type sensors in that a permanent magnet sealed inside a float travels up and down a stem in which a magnetostrictive wire is sealed. Ideal for high-accuracy, continuous level measurement of a wide variety of liquids in storage and shipping containers, these sensors require the proper choice of float based on the specific gravity of the liquid. When choosing float and stem materials for magnetostrictive level sensors, the same guidelines described for magnetic and mechanical float level sensors apply.

Magnetostrictive level and position devices charge the magnetostrictive wire with electric current, when the field intersects the floats' magnetic field a mechanical twist or pulse is generated, this travels back down the wire at the speed of sound, like ultrasound or radar the distance is measured by time of flight from pulse to return pulse registry. the time of flight corresponds to the distance from the sensor detecting the return pulse.

Because of the accuracy possible with the magnetostrictive technique, it is popular for "custody-transfer" applications. It can be permitted by an agency of weights and measures for conducting commercial transactions. It is also frequently applied on magnetic sight gages. In this variation, the magnet is installed in a float that travels inside a gage glass or tube. The magnet operates on the sensor which is mounted externally on the gage. Boilers and other high temperature or pressure applications take advantage of this performance quality

Resistive Chain

Resistive chain level sensors are similar to magnetic float level sensors in that a permanent magnet sealed inside a float moves up and down a stem in which closely spaced switches and resistors are sealed. When the switches are closed, the resistance is summed and converted to current or voltage signals that are proportional to the level of the liquid.

The choice of float and stem materials depends on the liquid in terms of chemical compatibility as well as specific gravity and other factors that affect buoyancy. These sensors work well for liquid level measurements in marine, chemical processing, pharmaceuticals, food processing, waste treatment, and other applications. With the proper choice of two floats, resistive chain level sensors can also be used to monitor for the presence of an interface between two immiscible liquids whose specific gravities are more than 0.6, but differ by as little as 0.1 unit.

Magnetoresistive

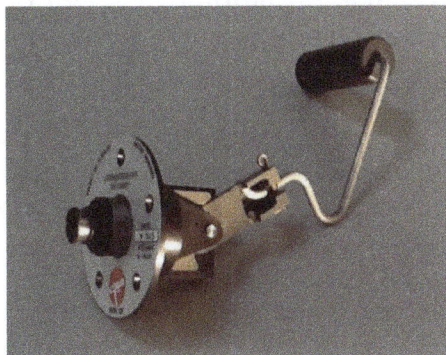

Magneto Resistive Level Sensor

Magnetoresistance float level sensors are similar to float level sensors however a permanent magnet pair is sealed inside the float arm pivot. As the float moves up the motion and location are transmitted as the angular position of the magnetic field. This detection system is highly accurate down to 0.02° of motion. The field compass location provides a physical location of the float position. The choice of float and stem materials depends on the liquid in terms of chemical compatibility as well as specific gravity and other factors that affect buoyancy of the float. The electronic monitoring system does not come in contact with the fluid and is considered Intrinsic safety: or explosion proof. These sensors work well for liquid level measurements in marine, vehicle, aviation, chemical processing, pharmaceuticals, food processing, waste treatment, and other applications.

Due to the presence of a microprocessor and low power consumption, there is also capability for serial communication from to other computing devices making this a good technique for adjusting calibration and filtering of the sensor signal.

Hydrostatic Pressure

Hydrostatic pressure level sensors are submersible or externally mounted pressure sensors suitable for measuring the level of corrosive liquids in deep tanks or water in reservoirs. For these sensors, using chemically compatible materials is important to assure proper performance. Sensors are commercially available from 10mbar to 1000bar.

Since these sensors sense increasing pressure with depth and because the specific gravities of liquids are different, the sensor must be properly calibrated for each application. In addition, large variations in temperature cause changes in specific gravity that should be accounted for when the pressure is converted to level. These sensors can be designed to keep the diaphragm free of contamination or build-up, thus ensuring proper operation and accurate hydrostatic pressure level measurements.

For use in open air applications, where the sensor cannot be mounted to the bottom of the tank or pipe thereof, a special version of the hydrostatic pressure level sensor can be suspended from a cable into the tank to the bottom point that is to be measured. The sensor must be specially designed to seal the electronics from the liquid environment. In tanks with a small head pressure (less than 100 INWC), it is very important to vent the back of the sensor gauge to atmospheric pressure. Otherwise, normal changes in barometric pressure will introduce large error in the sensor output signal. In addition, most sensors need to be compensated for temperature changes in the fluid..

Air Bubbler

An air bubbler system uses a tube with an opening below the surface of the liquid level. A fixed flow of air is passed through the tube. Pressure in the tube is proportional to the depth (and density) of the liquid over the outlet of the tube.

Air bubbler systems contain no moving parts, making them suitable for measuring the level of sewage, drainage water, sewage sludge, night soil, or water with large quantities of suspended solids. The only part of the sensor that contacts the liquid is a bubble tube which is chemically compatible with the material whose level is to be measured. Since the point of measurement has no electrical components, the technique is a good choice for classified "Hazardous Areas". The control portion of the system can be located safely away, with the pneumatic plumbing isolating the hazardous from the safe area.

Air bubbler systems are a good choice for open tanks at atmospheric pressure and can be built so that high-pressure air is routed through a bypass valve to dislodge solids that may clog the bubble tube. The technique is inherently "self-cleaning". It is highly recommended for liquid level measurement applications where ultrasonic, float or microwave techniques have proved undependable. The system will require constant supply of air during measurement. The end of the tube should be above certain height to avoid sludge from clogging the tube.

Gamma Ray

A nuclear level gauge or gamma ray gauge measures level by the attenuation of gamma rays passing through a process vessel. The technique is used to regulate the level of molten steel in a continuous casting process of steelmaking. The water-cooled mold is arranged with a source of radiation, such as cobalt-60 or caesium-137, on one side and a sensitive detector such as a scintillation counter on the other. As the level of molten steel rises in the mold, less of the gamma radiation is detected by the sensor. The technique allows non-contact measurement where the heat of the molten metal makes contact techniques and even many non-contact techniques impractical.

Biosensor

A biosensor is an analytical device, used for the detection of an analyte, that combines a biological component with a physicochemical detector. The *sensitive biological element* (e.g. tissue, microorganisms, organelles, cell receptors, enzymes, antibodies, nucleic acids, etc.) is a biologically derived material or biomimetic component that interacts (binds or recognizes) with the analyte under study. The biologically sensitive elements can also be created by biological engineering. The *transducer* or the *detector element* (works in a physicochemical way; optical, piezoelectric, electrochemical, etc.) transforms the signal resulting from the interaction of the analyte with the biological element into another signal (i.e., transduces) that can be more easily measured and quantified. The biosensor reader device with the associated electronics or signal processors that are primarily responsible for the display of the results in a user-friendly way. This sometimes accounts for the most expensive part of the sensor device, however it is possible to generate a user friendly display that includes transducer and sensitive element (holographic sensor). The readers are usually custom-designed and manufactured to suit the different working principles of biosensors.

Biosensor System

A biosensor typically consists of a bio-recognition component, biotransducer component, and electronic system which include a signal amplifier, processor, and display. Transducers and electronics can be combined, e.g., in CMOS-based microsensor systems. The recognition component, often called a bioreceptor, uses biomolecules from organisms or receptors modeled after biological systems to interact with the analyte of interest. This interaction is measured by the biotransducer which outputs a measurable signal proportional to the presence of the target analyte in the sample. The general aim of the design of a biosensor is to enable quick, convenient testing at the point of concern or care where the sample was procured.

Bioreceptors

In a biosensor, the bioreceptor is designed to interact with the specific analyte of interest to produce an effect measurable by the transducer. High selectivity for the analyte among a matrix of other chemical or biological components is a key requirement of the bioreceptor. While the type of biomolecule used can vary widely, biosensors can be classified according to common types bioreceptor interactions involving: anitbody/antigen, enzymes, nucleic acids/DNA, cellular structures/cells, or biomimetic materials.

Antibody/Antigen Interactions

An immunosensor utilizes the very specific binding affinity of antibodies for a specific compound or antigen. The specific nature of the antibody antigen interaction is analogous to a lock and key fit in that the antigen will only bind to the antibody if it has the correct conformation. Binding events result in a physicochemical change that in combination with a tracer, such as a fluorescent molecules, enzymes, or radioisotopes, can generate a signal. There are limitations with using antibodies in sensors: 1.The antibody binding capacity is strongly dependent on assay conditions (e.g. pH and temperature) and 2. The antibody-antigen interaction is generally irreversible. However, it has been shown that binding can be disrupted by chaotropic reagents, organic solvents, or even ultrasonic radiation.

Enzymatic Interactions

The specific binding capabilities and catalytic activity of enzymes make them popular bioreceptors. Analyte recognition is enabled through several possible mechanisms: 1) the enzyme converting the analyte into a product that is sensor-detectable, 2) detecting enzyme inhibition or activation by the analyte, or 3) monitoring modification of enzyme properties resulting from interaction with the analyte. The main reasons for the common use of enzymes in biosensors are: 1) ability to catalyze a large number of reactions; 2) potential to detect a group of analytes (substrates, products, inhibitors, and modulators of the catalytic activity); and 3) suitability with several different transduction methods for detecting the analyte. Notably, since enzymes are not consumed in reactions, the biosensor can easily be used continuously. The catalytic activity of enzymes also allows lower limits of detection compared to common binding techniques. However, the sensor's lifetime is limited by the stability of the enzyme.

Nucleic Acid Interactions

Biosensors that employ nucleic acid interactions can be referred to as genosensors. The recognition process is based on the principle of complementary base pairing, adenine:thymine and cytosine:guanine in DNA. If the target nucleic acid sequence is known,complementary sequences can be synthesized, labeled, and then immobilized on the sensor. The hybridization probes can then base pair with the target sequences, generating an optical signal. The favored transduction principle employed in this type of sensor has been optical detection.

Epigenetics

It has been proposed that properly optimized integrated optical resonators can be exploited for detecting epigenetic modifications (e.g. DNA methylation, histone post-translational modifications) in body fluids from patients affected by cancer or other diseases. Photonic biosensors with ultra-sensitivity are nowadays being developed at a research level to easily detect cancerous cells within the patient's urine. Different research projects aim to develop new portable devices that uses cheap, environmentally friendly, disposable cartridges that require only simple handling with no need of further processing, washing, or manipulation by expert technicians.

Organelles

Organelles form separate compartments inside cells and usually perform function independently. Different kinds of organelles have various metabolic pathways and contain enzymes to fulfill its function. Commonly used organelles include lysosome, chloroplast and mitochondria. The spatial-temporal distribution pattern of calcium is closed related to ubiquitous signaling pathway. Mitochondria actively participate in the metabolism of calcium ions to control the function and also modulate the calcium related signaling pathways. Experiments have proved that mitochondria have the ability to respond to high calcium concentration generated in the proximity by opening the calcium channel. In this way, mitochondria can be used to detect the calcium concentration in medium and the detection is very sensitive due to high spatial resolution. Another application of mitochondria is used for detection of water pollution. Detergent compounds' toxicity will damage the cell and subcellular structure including mitochondria. The detergents will cause a swelling effect which could be measured by an absorbance change. Experiment data shows the change rate is proportional to the detergent concentration, providing a high standard for detection accuracy.

Cells

Cells are often used in bioreceptors because they are sensitive to surrounding environment and they can respond to all kinds of stimulants. Cells tend to attach to the surface so they can be easily immobilized. Compared to organelles they remain active for longer period and the reproducibility makes them reusable. They are commonly used to detect global parameter like stress condition, toxicity and organic derivatives. They can also be used to monitor the treatment effect of drugs. One application is to use cells to determine herbicides which are main aquatic contaminant. Microalgae are entrapped on a quartz microfiber and the chlorophyll fluorescence modified by herbicides is collected at the tip of an optical fiber bundle and transmitted to a fluorimeter. The algae are continuously cultured to get optimized measurement. Results show that detection limit of certain herbicide can reach sub-ppb concentration level. Some cells can also be used to monitor the microbial corrosion. Pseudomonas sp. is isolated form corroded material surface and immobilized on acetylcellulose membrane. The respiration activity is determined by measuring oxygen consumption. There is linear relationship between the current generated and the concentration of sulfuric acid. The response time is related to the loading of cells and surrounding environments and can be controlled to no more than 5min.

Tissue

Tissues are used for biosensor for the abundance of enzymes existed. Advantages of tissues as biosensors include the following: 1)easier to immobilize compared to cells and organelles 2)the higher activity and stability from maintain enzymes in natural environment 3)the availability and low price 4)the avoidance of tedious work of extraction, centrifuge and purification of enzymes 5)necessary cofactors for enzyme to function exists 6)the diversity providing a wide range of choice concerning different objectives. There also exists some disadvantages of tissues like the lack of specificity due to the interference of other enzymes and longer response time due to transport barrier.

Surface Attachment of The Biological Elements

An important part in a biosensor is to attach the biological elements (small molecules/protein/cells) to the surface of the sensor (be it metal, polymer or glass). The simplest way is to functionalize the surface in order to coat it with the biological elements. This can be done by polylysine, aminosilane, epoxysilane or nitrocellulose in the case of silicon chips/silica glass. Subsequently the bound biological agent may be for example fixed by Layer by layer deposition of alternatively charged polymer coatings Alternatively three-dimensional lattices (hydrogel/xerogel) can be used to chemically or physically entrap these (where by chemically entraped it is meant that the biological element is kept in place by a strong bond, while physically they are kept in place being unable to pass through the pores of the gel matrix). The most commonly used hydrogel is sol-gel, a glassy silica generated by polymerization of silicate monomers (added as tetra alkyl orthosilicates, such as TMOS or TEOS) in the presence of the biological elements (along with other stabilizing polymers, such as PEG) in the case of physical entrapment. Another group of hydrogels, which set under conditions suitable for cells or protein, are acrylate hydrogel, which polymerize upon radical initiation. One type of radical initiator is a peroxide radical, typically generated by combining a persulfate with TEMED (Polyacrylamide gel are also commonly used for protein electrophoresis), alternatively light can be used in combination with a photoinitiator, such as DMPA (2,2-dimethoxy-2-phenylacetophenone). Smart materials that mimic the biological components of a sensor can also be classified as biosensors using only the active or catalytic site or analogous configurations of a biomolecule.

Biotransducer

Biosensors can be classified by their biotransducer type. The most common types of biotransducers used in biosensors are 1) electrochemical biosensors, 2) optical biosensors, 3)electronic biosensors, 4)piezoelectric biosensors, 5) gravimetric biosensors, 6) pyroelectric biosensors.

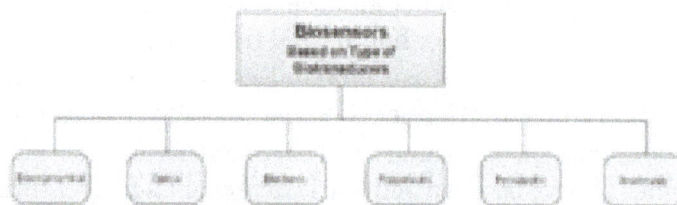

Classification of Biosensors based on type of biotransducer

Electrochemical

Electrochemical biosensors are normally based on enzymatic catalysis of a reaction that produces or consumes electrons (such enzymes are rightly called redox enzymes). The sensor substrate usually contains three electrodes; a reference electrode, a working electrode and a counter electrode. The target analyte is involved in the reaction that takes place on the active electrode surface, and the reaction may cause either electron transfer across the double layer (producing a current) or can contribute to the double layer potential (producing a voltage). We can either measure the current (rate of flow of electrons is now proportional to the analyte concentration) at a fixed potential or the potential can be measured at zero current (this gives a logarithmic response). Note that potential of the working or active electrode is space charge sensitive and this is often used. Further, the label-free and direct electrical detection of small peptides and proteins is possible by their intrinsic charges using biofunctionalized ion-sensitive field-effect transistors.

Another example, the potentiometric biosensor, (potential produced at zero current) gives a logarithmic response with a high dynamic range. Such biosensors are often made by screen printing the electrode patterns on a plastic substrate, coated with a conducting polymer and then some protein (enzyme or antibody) is attached. They have only two electrodes and are extremely sensitive and robust. They enable the detection of analytes at levels previously only achievable by HPLC and LC/MS and without rigorous sample preparation. All biosensors usually involve minimal sample preparation as the biological sensing component is highly selective for the analyte concerned. The signal is produced by electrochemical and physical changes in the conducting polymer layer due to changes occurring at the surface of the sensor. Such changes can be attributed to ionic strength, pH, hydration and redox reactions, the latter due to the enzyme label turning over a substrate. Field effect transistors, in which the gate region has been modified with an enzyme or antibody, can also detect very low concentrations of various analytes as the binding of the analyte to the gate region of the FET cause a change in the drain-source current.

Ion Channel Switch

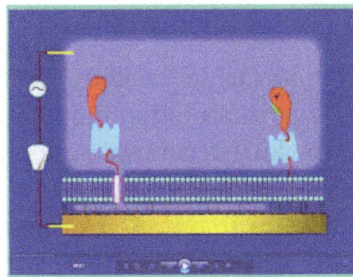

ICS – channel open

The use of ion channels has been shown to offer highly sensitive detection of target biological molecules. By embedding the ion channels in supported or tethered bilayer membranes (t-BLM) attached to a gold electrode, an electrical circuit is created. Capture molecules such as antibodies can be bound to the ion channel so that the binding of the target molecule controls the ion flow through the channel. This results in a measurable change in the electrical conduction which is proportional to the concentration of the target.

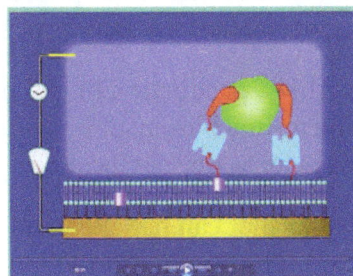

ICS – channel closed

An ion channel switch (ICS) biosensor can be created using gramicidin, a dimeric peptide channel, in a tethered bilayer membrane. One peptide of gramicidin, with attached antibody, is mobile and one is fixed. Breaking the dimer stops the ionic current through the membrane. The magnitude of the change in electrical signal is greatly increased by separating the membrane from the metal surface using a hydrophilic spacer.

Quantitative detection of an extensive class of target species, including proteins, bacteria, drug and toxins has been demonstrated using different membrane and capture configurations.

Others

Piezoelectric sensors utilise crystals which undergo an elastic deformation when an electrical potential is applied to them. An alternating potential (A.C.) produces a standing wave in the crystal at a characteristic frequency. This frequency is highly dependent on the elastic properties of the crystal, such that if a crystal is coated with a biological recognition element the binding of a (large) target analyte to a receptor will produce a change in the resonance frequency, which gives a binding signal. In a mode that uses surface acoustic waves (SAW), the sensitivity is greatly increased. This is a specialised application of the Quartz crystal microbalance as a biosensor

Thermometric and Magnetic Based Biosensors are Rare.

Placement of Biosensors

An in-vivo biosensor is one that functions inside the body. Biocompatibility concerns follow with the creation of an in-vivo biosensor. That is, an initial inflammatory response occurring after the implantation. The second concern is the long-term interaction with the body during the intended period of the device's use. Another issue that arises is failure. If there is failure, the device must be removed and replaced, causing additional surgery. An example for application of an in-vivo biosensor is insulin monitoring within the body.

An in-vitro biosensor is a sensor that takes place in a test tube, culture dish, or elsewhere outside a living organism. The sensor uses a biological element, such as an enzyme capable of recognizing or signaling a biochemical change in solution. A transducer is then used to convert the biochemical signal to a quantifiable signal. An example of an in-vitro biosensor is an enzyme-conductimetric biosensor for glucose monitoring.

An at-line biosensor is used in a production line where a sample can be taken, tested, and a decision can be made whether or not the continuation of the production should occur. An example of an at-line biosensor is the monitoring of lactose in a dairy processing plant.

The in line biosensor can be placed within a production line to monitor a variable with continuous production and can be automated. The in-line biosensor becomes another step in the process line. An application of an in-line biosensor is for water purification.

There is a challenge to create a biosensor that can be taken straight to the "point of concern", that is the location where the test is needed. The elimination of lab testing can save time and money. An application of a point-of-concern biosensor can be for the testing of HIV virus in third world countries where it is difficult for the patients to be tested. A biosensor can be sent directly to the location and a quick and easy test can be used.

Applications

There are many potential applications of biosensors of various types. The main requirements for a biosensor approach to be valuable in terms of research and commercial applications are the identification of a target molecule, availability of a suitable biological recognition element, and the potential for disposable portable detection systems to be preferred to sensitive laboratory-based techniques in some situations. Some examples are given below:

- Glucose monitoring in diabetes patients

- Other medical health related targets

- Environmental applications e.g. the detection of pesticides and river water contaminants such as heavy metal ions

- Remote sensing of airborne bacteria e.g. in counter-bioterrorist activities

- Remote sensing of water quality in coastal waters by describing online different aspects of clam ethology (biological rhythms, growth rates, spawning or death records) in groups of abandoned bivalves around the world

- Detection of pathogens

- Determining levels of toxic substances before and after bioremediation

- Detection and determining of organophosphate

- Routine analytical measurement of folic acid, biotin, vitamin B12 and pantothenic acid as an alternative to microbiological assay

- Determination of drug residues in food, such as antibiotics and growth promoters, particularly meat and honey.

- Drug discovery and evaluation of biological activity of new compounds.

- Protein engineering in biosensors

- Detection of toxic metabolites such as mycotoxins

A common example of a commercial biosensor is the blood glucose biosensor, which uses the enzyme glucose oxidase to break blood glucose down. In doing so it first oxidizes glucose and uses two electrons to reduce the FAD (a component of the enzyme) to FADH2. This in turn is oxidized by the electrode in a number of steps. The resulting current is a measure of the concentration of glucose. In this case, the electrode is the transducer and the enzyme is the biologically active component.

A canary in a cage, as used by miners to warn of gas, could be considered a biosensor. Many of today's biosensor applications are similar, in that they use organisms which respond to toxic substances at a much lower concentrations than humans can detect to warn of their presence. Such devices can be used in environmental monitoring, trace gas detection and in water treatment facilities.

Many optical biosensors are based on the phenomenon of surface plasmon resonance (SPR) techniques. This utilises a property of and other materials; specifically that a thin layer of gold on a high refractive index glass surface can absorb laser light, producing electron waves (surface plasmons) on the gold surface. This occurs only at a specific angle and wavelength of incident light and is highly dependent on the surface of the gold, such that binding of a target analyte to a receptor on the gold surface produces a measurable signal.

Surface plasmon resonance sensors operate using a sensor chip consisting of a plastic cassette supporting a glass plate, one side of which is coated with a microscopic layer of gold. This side contacts the optical detection apparatus of the instrument. The opposite side is then contacted with a microfluidic flow system. The contact with the flow system creates channels across which reagents can be passed in solution. This side of the glass sensor chip can be modified in a number of ways, to allow easy attachment of molecules of interest. Normally it is coated in carboxymethyl dextran or similar compound.

The refractive index at the flow side of the chip surface has a direct influence on the behavior of the light reflected off the gold side. Binding to the flow side of the chip has an effect on the refractive index and in this way biological interactions can be measured to a high degree of sensitivity with some sort of energy. The refractive index of the medium near the surface changes when biomolecules attach to the surface, and the SPR angle varies as a function of this change.

Light of a fixed wavelength is reflected off the gold side of the chip at the angle of total internal reflection, and detected inside the instrument. The angle of incident light is varied in order to match the evanescent wave propagation rate with the propagation rate of the surface plasmon plaritons. This induces the evanescent wave to penetrate through the glass plate and some distance into the liquid flowing over the surface.

Other optical biosensors are mainly based on changes in absorbance or fluorescence of an appropriate indicator compound and do not need a total internal reflection geometry. For example, a fully operational prototype device detecting casein in milk has been fabricated. The device is based on detecting changes in absorption of a gold layer. A widely used research tool, the micro-array, can also be considered a biosensor.

Biological biosensors often incorporate a genetically modified form of a native protein or enzyme. The protein is configured to detect a specific analyte and the ensuing signal is read by a detection instrument such as a fluorometer or luminometer. An example of a recently developed biosensor is one for detecting cytosolic concentration of the analyte cAMP (cyclic adenosine monophosphate), a second messenger involved in cellular signaling triggered by ligands interacting with receptors on the cell membrane. Similar systems have been created to study cellular responses to native ligands or xenobiotics (toxins or small molecule inhibitors). Such "assays" are commonly used in drug discovery development by pharmaceutical and biotechnology companies. Most cAMP assays in current use require lysis of the cells prior to measurement of cAMP. A live-cell biosensor for cAMP can be used in non-lysed cells with the additional advantage of multiple reads to study the kinetics of receptor response.

Nanobiosensors use an immobilized bioreceptor probe that is selective for target analyte molecules. Nanomaterials are exquisitely sensitive chemical and biological sensors. Nanoscale materials demonstrate unique properties. Their large surface area to volume ratio can achieve rapid and low cost reactions, using a variety of designs.

Other evanescent wave biosensors have been commercialised using waveguides where the propagation constant through the waveguide is changed by the absorption of molecules to the waveguide surface. One such example, dual polarisation interferometry uses a buried waveguide as a reference against which the change in propagation constant is measured. Other configurations such as the Mach–Zehnder have reference arms lithographically defined on a substrate. Higher levels of integration can be achieved using resonator geometries where the resonant frequency of a ring resonator changes when molecules are absorbed.

Recently, arrays of many different detector molecules have been applied in so called electronic nose devices, where the pattern of response from the detectors is used to fingerprint a substance. In the Wasp Hound odor-detector, the mechanical element is a video camera and the biological element is five parasitic wasps who have been conditioned to swarm in response to the presence of a specific chemical. Current commercial electronic noses, however, do not use biological elements.

Glucose Monitoring

Commercially available gluocose monitors rely on amperometric sensing of glucose by means of glucose oxidase, which oxidises glucose producing hydrogen peroxide which is detected by the electrode. To overcome the limitation of amperometric sensors, a flurry of research is present into novel sensing methods, such as fluorescent glucose biosensors.

Interferometric Reflectance Imaging Sensor

The interferometric reflectance imaging sensor (IRIS) is based on the principles of optical interference and consists of a silicon-silicon oxide substrate, standard optics, and low-powered coherent LEDs. When light is illuminated through a low magnification objective onto the layered silicon-silicon oxide substrate, an interferometric signature is produced. As biomass, which has a similar index of refraction as silicon oxide, accumulates on the substrate surface, a change in the interferometric signature occurs and the change can be correlated to a quantifiable mass. *Daaboul et al* used IRIS to yield a label-free sensitivity of approximately 19 ng/mL. *Ahn et al.* improved the sensitivity of IRIS through a mass tagging technique.

Since initial publication, IRIS has been adapted to perform various functions. First, IRIS integrated a fluorescence imaging capability into the interferometric imaging instrument as a potential way to address fluorescence protein microarray variability. Briefly, the variation in fluorescence microarrays mainly derives from inconsistent protein immobilization on surfaces and may cause misdiagnoses in allergy microarrays. To correct from any variation in protein immobilization, data acquired in the fluorescence modality is then normalized by the data acquired in the label-free modality. IRIS has also been adapted to perform single nanoparticle counting by simply switching the low magnification objective used for label-free biomass quantification to a higher objective magnification. This modality enables size discrimination in complex human biological samples. *Monroe et al.* used IRIS to quantify protein levels spiked into human whole blood and serum and determined allergen sensitization in characterized human blood samples using zero sample processing. Other practical uses of this device include virus and pathogen detection.

Food Analysis

There are several applications of biosensors in food analysis. In the food industry, optics coated with antibodies are commonly used to detect pathogens and food toxins. Commonly, the light system in these biosensors is fluorescence, since this type of optical measurement can greatly amplify the signal.

A range of immuno- and ligand-binding assays for the detection and measurement of small molecules such as water-soluble vitamins and chemical contaminants (drug residues) such as sulfonamides and Beta-agonists have been developed for use on SPR based sensor systems, often adapted from existing ELISA or other immunological assay. These are in widespread use across the food industry.

DNA Biosensors

In the future, DNA will find use as a versatile material from which scientists can craft biosensors. DNA biosensors can theoretically be used for medical diagnostics, forensic science, agriculture, or even environmental clean-up efforts. No external monitoring is needed for DNA-based sensing devises. This is a significant advantage. DNA biosensors are complicated mini-machines—consisting of sensing elements, micro lasers, and a signal generator. At the heart of DNA biosensor function is the fact that two strands of DNA stick to each other by virtue of chemical attractive forces. On such a sensor, only an exact fit—that is, two strands that match up at every nucleotide position—gives rise to a fluorescent signal (a glow) that is then transmitted to a signal generator.

Microbial Biosensors

Using biological engineering researchers have created many microbial biosensors. An example is the arsenic biosensor. To detect arsenic they use the Ars operon. Using bacteria, researchers can detect pollutants in samples.

Ozone Biosensors

because ozone filters out harmful ultraviolet radiation, the discovery of holes in the ozone layer of the earth's atmosphere has raised concern about how much ultraviolet light reaches the earth's surface. Of particular concern are the questions of how deeply into sea water ultraviolet radiation penetrates and how it affects marine organisms, especially plankton (floating microorganisms) and viruses that attack plankton. Plankton form the base of the marine food chains and are believed to affect our planet's temperature and weather by uptake of CO_2 for photosynthesis.

Deneb Karentz, a researcher at the Laboratory of Radio-biology and Environmental Health (University of California in San Francisco) has devised a simple method for measuring ultraviolet penetration and intensity. Working in the Antarctic Ocean, she submerfed to various depths thin plastic bags containing special strains of *E. coli* that are almost totally unable to repair ultraviolet radiation damage to their DNA. Bacterial death rates in these bags were compared with rates in unexposed control bags of the same organism. The bacterial "biosensors" revealed constant significant ultraviolet damage at depths of 10 m and frequently at 20 and 30 m. Karentz plans additional studies of how ultraviolet may affect seasonal plankton blooms (growth spurts) in the oceans.

Metastatic Cancer Cell Biosensors

Metastasis is the spread of cancer from one part of the body to another via either the circulatory system or lymphatic system. Unlike radiology imaging tests (mammograms), which send forms of energy (x-rays, magnetic fields, etc.) through the body to only take interior pictures, biosensors have the potential to directly test the malignant power of the tumor. The combination of a biological and detector element allows for a small sample requirement, a compact design, rapid signals, rapid detection, high selectivity and high sensitivity for the analyte being studied. Compared to the usual radiology imaging tests biosensors have the advantage of not only finding out how far the cancer has spread and checking if treatment is effective, but also are cheaper, more efficient (in time, cost and productivity) ways to assess metastaticity in early stages of cancer. Biological engineering researchers have created oncological biosensors for breast cancer. Breast cancer is the leading common cancer among women worldwide. An example would be a Transferrin- Quartz crystal microbalance (QCM). As a biosensor, quartz crystal microbalances produce oscillations in the frequency of the crystal's standing wave from an alternating potential to detect nano-gram mass changes. These biosensors are specifically designed to interact and have high selectivity for receptors on cell (cancerous and normal) surfaces. Ideally this provides a quantitative detection of cells with this receptor per surface area instead of a qualitative picture detection given by mammograms.

Seda Atay, a Biotechnology researcher at Hacettepe University, experimentally observed this specificity and selectivity between a QCM and MDA-MB 231 breast cells, MCF 7 cells, and starved MDA-MB 231 cells in vitro. With other researchers she devised a method of washing these different metastatic leveled cells over the sensors to measure mass shifts due to different quantities of transferrin receptors. Particularly, the metastatic power of Breast cancer cells can be determined by Quartz crystal microbalances with nanoparticles and transferrin that would potentially attach to transferrin receptors on cancer cell surfaces. There is very high selectivity for transferrin receptors because they are over-expressed in cancer cells. If cells have high expression of transferrin receptors, which shows their high metastatic power, they have higher affinity and bind more to the QCM that measures the increase in mass. Depending on the magnitude of the nano-gram mass change, the metastatic power can be determined.

The Purpose of Nanoparticles

When nanoparticles are attached to the QCM surface their simplicity, variability in shape, high surface area, physicochemical malleability, and optional attachment of metals enables for different properties, a change in responses, selectivities and specificities. This transducer's characteristics in combination with nanoparticles with large surface area to volume ratios makes it a perfect biosensor to particularly determine the metastatic power and malignancy of cancer cells. In Seda Atay's study, determining the metastatic power of in vitro breast cancer exactly required 58 nm sized Poly(2-hydroxyethyl methacrylate) (PHEMA) nanoparticles with a surface area of 1899 m2g-1 to effectively adsorb the cells to the QCM surface.

Blood Glucose Monitoring

Blood glucose monitoring is a way of testing the concentration of glucose in the blood (glycemia). Particularly important in the care of diabetes mellitus, a blood glucose test is performed by piercing the skin (typically, on the finger) to draw blood, then applying the blood to a chemically active disposable 'test-strip'. Different manufacturers use different technology, but most systems measure an electrical characteristic, and use this to determine the glucose level in the blood. The test is usually referred to as capillary blood glucose.

Healthcare professionals advise patients with diabetes on the appropriate monitoring regimen for their condition. Most people with Type 2 diabetes test at least once per day. Diabetics who use insulin (all Type 1 diabetes and many Type 2s) usually test their blood sugar more often (3 to 10 times per day), both to assess the effectiveness of their prior insulin dose and to help determine their next insulin dose.

Improved technology for measuring blood glucose is rapidly changing the standards of care for all diabetic people.

Purpose

Blood glucose monitoring reveals individual patterns of blood glucose changes, and helps in the planning of meals, activities, and at what time of day to take medications.

Also, testing allows for quick response to high blood sugar (hyperglycemia) or low blood sugar (hypoglycemia). This might include diet adjustments, exercise, and insulin (as instructed by the health care provider).

Blood Glucose Meters

A blood glucose meter is an electronic device for measuring the blood glucose level. A relatively small drop of blood is placed on a disposable test strip which interfaces with a digital meter. Within several seconds, the level of blood glucose will be shown on the digital display.

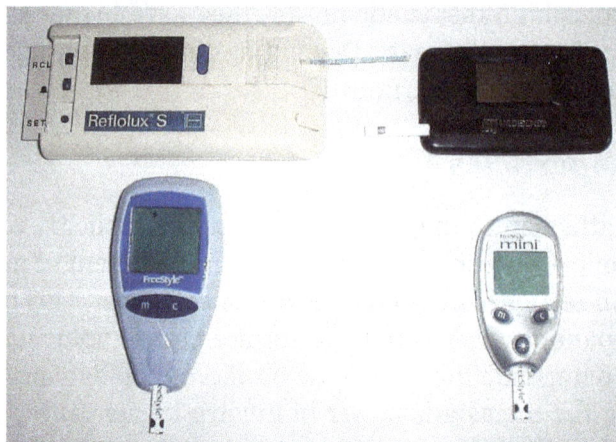

Four generations of blood glucose meter, c. 1991–2005. Sample sizes vary from 30 to 0.3 µl. Test times vary from 5 seconds to 2 minutes (modern meters are typically below 15 seconds).

Needing only a small drop of blood for the meter means that the time and effort required for testing is reduced and the compliance of diabetic people to their testing regimens is improved. Although the cost of using blood glucose meters seems high, it is believed to be a cost benefit relative to the avoided medical costs of the complications of diabetes.

Recent advances include:

- 'alternate site testing', the use of blood drops from places other than the finger, usually the palm or forearm. This alternate site testing uses the same test strips and meter, is practically pain free, and gives the real estate on the finger tips a needed break if they become sore. The disadvantage of this technique is that there is usually less blood flow to alternate sites, which prevents the reading from being accurate when the blood sugar level is changing.

- 'no coding' systems. Older systems required 'coding' of the strips to the meter. This carried a risk of 'miscoding', which can lead to inaccurate results. Two approaches have resulted in systems that no longer require coding. Some systems are 'autocoded', where technology is used to code each strip to the meter. And some are manufactured to a 'single code', thereby avoiding the risk of miscoding.

- 'multi-test' systems. Some systems use a cartridge or a disc containing multiple test strips. This has the advantage that the user doesn't have to load individual strips each time, which is convenient and can enable quicker testing.

- 'downloadable' meters. Most newer systems come with software that allows the user to download meter results to a computer. This information can then be used, together with health care professional guidance, to enhance and improve diabetes management. The meters usually require a connection cable, unless they are designed to work wirelessly with an insulin pump, or are designed to plug directly into the computer.

Continuous Glucose Monitoring

A continuous glucose monitor (CGM) determines glucose levels on a continuous basis (every few minutes). A typical system consists of:

- a disposable glucose sensor placed just under the skin, which is worn for a few days until replacement

- a link from the sensor to a non-implanted transmitter which communicates to a radio receiver

- an electronic receiver worn like a pager (or insulin pump) that displays glucose levels with nearly continuous updates, as well as monitors rising and falling trends.

Continuous glucose monitors measure the glucose level of interstitial fluid. Shortcomings of CGM systems due to this fact are:

- continuous systems must be calibrated with a traditional blood glucose measurement (using current technology) and therefore require both the CGM system and occasional "fingerstick"

- glucose levels in interstitial fluid lag behind blood glucose values

Patients therefore require traditional fingerstick measurements for calibration (typically twice per day) and are often advised to use fingerstick measurements to confirm hypo- or hyperglycemia before taking corrective action.

The lag time discussed above has been reported to be about 5 minutes. Anecdotally, some users of the various systems report lag times of up to 10–15 minutes. This lag time is insignificant when blood sugar levels are relatively consistent. However, blood sugar levels, when changing rapidly, may read in the normal range on a CGM system while in reality the patient is already experiencing symptoms of an out-of-range blood glucose value and may require treatment. Patients using CGM are therefore advised to consider both the absolute value of the blood glucose level given by the system as well as any trend in the blood glucose levels. For example, a patient using CGM with a blood glucose of 100 mg/dl on their CGM system might take no action if their blood glucose has been consistent for several readings, while a patient with the same blood glucose level but whose blood glucose has been dropping steeply in a short period of time might be advised to perform a fingerstick test to check for hypoglycemia.

Continuous monitoring allows examination of how the blood glucose level reacts to insulin, exercise, food, and other factors. The additional data can be useful for setting correct insulin dosing ratios for food intake and correction of hyperglycemia. Monitoring during periods when blood glucose levels are not typically checked (e.g. overnight) can help to identify problems in insulin dosing (such as basal levels for insulin pump users or long-acting insulin levels for patients taking injections). Monitors may also be equipped with alarms to alert patients of hyperglycemia or hypoglycemia so that a patient can take corrective action(s) (after fingerstick testing, if necessary) even in cases where they do not feel symptoms of either condition. While the technology has its limitations, studies have demonstrated that patients with continuous sensors experience less hyperglycemia and also reduce their glycosylated hemoglobin levels.

Currently, continuous blood glucose monitoring is not automatically covered by health insurance in the United States in the same way that most other diabetic supplies are covered (e.g. standard glucose testing supplies, insulin, and even insulin pumps). However, an increasing number of insurance companies do cover continuous glucose monitoring supplies (both the receiver and disposable sensors) on a case-by-case basis if the patient and doctor show a specific need. The lack of insurance coverage is exacerbated by the fact that disposable sensors must be frequently replaced. Some sensors have been U.S. Food and Drug Administration (FDA) approved for 7- and 3-day use, though some patients wear sensors for longer than the recommended period) and the receiving meters likewise have finite lifetimes (less than 2 years and as little as 6 months). This is one factor in the slow uptake in the use of sensors that have been marketed in the United States.

The principles, history and recent developments of operation of electrochemical glucose biosensors are discussed in a chemical review by Joseph Wang.

Some projects such Nightscout (DIY) allow parents to have a permanent control of glucose levels on their children from devices such cellphones, tablets, laptops, pebble watch, etc.

Glucose Sensing Bio-implants

Investigations on the use of test strips have shown that the required self-injury acts as a psychological barrier restraining the patients from sufficient glucose control. As a result, secondary diseases are caused by excessive glucose levels. A significant improvement of diabetes therapy might be achieved with an implantable sensor that would continuously monitor blood sugar levels within the body and transmit the measured data outside. The burden of regular blood testing would be taken from the patient, who would instead follow the course of their glucose levels on an intelligent device like a laptop or a smart phone.

Glucose concentrations do not necessarily have to be measured in blood vessels, but may also be determined in the interstitial fluid, where the same levels prevail – with a time lag of a few minutes – due to its connection with the capillary system. However, the enzymatic glucose detection scheme used in single-use test strips is not directly suitable for implants. One main problem is caused by the varying supply of oxygen, by which glucose is converted to glucono lactone and H_2O_2 by glucose oxidase. Since the implantation of a sensor into the body is accompanied by growth of encapsulation tissue, the diffusion of oxygen to the reaction zone is continuously diminished. This decreasing oxygen availability causes the sensor reading to drift, requiring frequent re-calibration using finger-sticks and test strips.

One approach to achieving long-term glucose sensing is to measure and compensate for the changing local oxygen concentration. Other approaches replace the troublesome glucose oxidase reaction with a reversible sensing reaction, known as an affinity assay. This scheme was originally put forward by Schultz & Sims in 1978. A number of different affinity assays have been investigated, with fluorescent assays proving most common. MEMS technology has recently allowed for smaller and more convenient alternatives to fluorescent detection, via measurement of viscosity. Investigation of affinity-based sensors has shown that encapsulation by body tissue does not cause a drift of the sensor signal, but only a time lag of the signal compared to the direct measurement in blood.

Non-invasive Technologies

Some new technologies to monitor blood glucose levels will not require access to blood to read the glucose level. Non-invasive technologies include near IR detection, ultrasound and dielectric spectroscopy. These may free the person with diabetes from finger sticks to supply the drop of blood for blood glucose analysis.

Most of the non-invasive methods under development are continuous glucose monitoring methods and offer the advantage of providing additional information to the subject between the conventional finger stick, blood glucose measurements and over time periods where no finger stick measurements are available (i.e. while the subject is sleeping).

Effectiveness

For patients with diabetes mellitus type 2, the importance of monitoring and the optimal frequency of monitoring are not clear. A 2011 study found no evidence that blood glucose monitoring leads to better patient outcomes in actual practice. One randomized controlled trial found that self-monitoring of blood glucose did not improve glycosylated hemoglobin (HbA1c) among "reasonably well controlled non-insulin treated patients with type 2 diabetes". However a recent meta-analysis of 47 randomized controlled trials encompassing 7677 patients showed that self-care management intervention improves glycemic control in diabetics, with an estimated 0.36% (95% CI, 0.21-0.51) reduction in their glycosylated hemoglobin values. Furthermore, a recent study showed that patients described as being "Uncontrolled Diabetics" (defined in this study by HbA1C levels >8%) showed a statistically significant decrease in the HbA1C levels after a 90-day period of seven-point self-monitoring of blood glucose (SMBG) with a relative risk reduction (RRR) of 0.18% (95% CI, 0.86-2.64%, p<.001). Regardless of lab values or other numerical parameters, the purpose of the clinician is to improve quality of life and patient outcomes in diabetic patients. A recent study included 12 randomized controlled trials and evaluated outcomes in 3259 patients. The authors concluded through a qualitative analysis that SMBG on quality of life showed no effect on patient satisfaction or the patients' health-related quality of life. Furthermore, the same study identified that patients with type 2 diabetes mellitus diagnosed greater than one year prior to initiation of SMBG, who were not on insulin, experienced a statistically significant reduction in their HbA1C of 0.3% (95% CI, -0.4 - -0.1) at six months follow up, but a statistically insignificant reduction of 0.1% (95% CI, -0.3 – 0.04) at twelve months follow up. Conversely, newly diagnosed patients experienced a statistically significant reduction of 0.5% (95% CI, -0.9 – -0.1) at 12 months follow up. A recent study found that a treatment strategy of intensively lowering blood sugar levels (below 6%) in patients with additional cardiovascular disease risk factors poses more harm than benefit. For type 2 diabetics who are not on insulin, exercise and diet are the best tools. Blood glucose monitoring is, in that case, simply a tool to evaluate the success of diet and exercise. Insulin-dependent type 2 diabetics do not need to monitor their blood sugar as frequently as type 1 diabetics.

Blood Glucose Monitoring Recommendations

The National Institute for Health and Clinical Excellence (NICE), UK released updated diabetes recommendations on 30 May 2008, which recommend that self-monitoring of plasma glucose levels for people with newly diagnosed type 2 diabetes must be integrated into a structured self-management education process. The recommendations have been updated in August 2015 for children and young adults with type 1 diabetes.

Load Cell

A load cell is a transducer that is used to create an electrical signal whose magnitude is directly proportional to the force being measured. The various types of load cells include hydraulic load cells, pneumatic load cells and strain gauge load cells.

Strain Gauge Load Cell

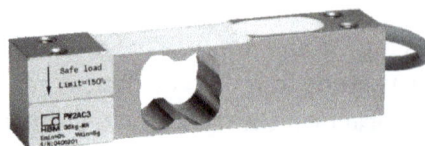

Double bending beam load cell element

Compression load cell

Through a mechanical construction, the force being sensed deforms a strain gauge. The strain gauge measures the deformation (strain) as a change in electrical resistance, which is a measure of the strain and hence the applied forces. A load cell usually consists of four strain gauges in a Wheatstone bridge configuration. Load cells of one strain gauge (quarter bridge) or two strain gauges (half bridge) are also available. The electrical signal output is typically in the order of a few millivolts and requires amplification by an instrumentation amplifier before it can be used. The output of the transducer can be scaled to calculate the force applied to the transducer. Sometimes a high resolution ADC, typically 24-bit, can be used directly.

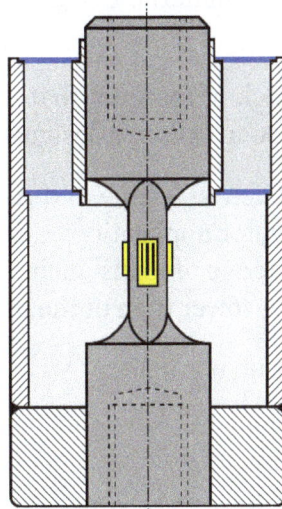

Push-pull rod load cell spring element

Strain gauge load cells are the most common in industry. These load cells are particularly stiff, have very good resonance values, and tend to have long life cycles in application. Strain gauge load cells work on the principle that the strain gauge (a planar resistor) deforms/stretches/contracts when the material of the load cells deforms appropriately. These values are extremely small and are relational to the stress and/or strain that the material load cell is undergoing at the time. The change in resistance of the strain gauge provides an electrical value change that is calibrated to the load placed on the load cell.

Strain gauge load cells convert the load acting on them into electrical signals. The gauges themselves are bonded onto a beam or structural member that deforms when weight is applied. In most cases, four strain gauges are used to obtain maximum sensitivity and temperature compensation. Two of the gauges are usually in tension can be represented as T1 and T2,and two in compression can be represented as C1 and C2, and are wired with compensation adjustments. The strain gauge load cell is fundamentally a spring optimized for strain measurement. Gauges are mounted in areas that exhibit strain in compression or tension. When weight is applied to the load cell, gauges C1 and C2 compress decreasing their resistances. Simultaneously, gauges T1 and T2 are stretched increasing their resistances. The change in resistances causes more current to flow through C1 and C2 and less current to flow through T1 and T2. Thus a potential difference is felt between the output or signal leads of the load cell. The gauges are mounted in a differential bridge to enhance measurement accuracy. When weight is applied, the strain changes the electrical resistance of the gauges in proportion to the load. Other load cells are fading into obscurity, as strain gauge load cells continue to increase their accuracy and lower their unit costs.

Common Shapes

There are several common shapes of load cells:

- Shear beam, a straight block of material fixed on one end and loaded on the other

- Double-ended shear beam, a straight block of material fixed at both ends and loaded in the center

- Compression load cell, a block of material designed to be loaded at one point or area in compression

- S-type load cell, a S-shaped block of material that can be used in both compression and tension (load links and tension load cells are designed for tension only)

- Rope clamp, an assembly attached to a rope and measures its tension. Rope clamps are popular in hoist, crane and elevator applications due to the ease of their installation; they have to be designed for a large range of loads, including dynamic peak loads, so their output for the rated load tends to be lower than of the other types

- Loadpin, used for sensing loads on e.g. axles

Common Issues

- Mechanical mounting: the cells have to be properly mounted. All the load force has to go through the part of the load cell where its deformation is sensed. Friction may induce offset or hysteresis. Wrong mounting may result in the cell reporting forces along undesired axis, which still may somewhat correlate to the sensed load, confusing the technician.

- Overload: Within its rating, the load cell deforms elastically and returns back to its shape after being unloaded. If subjected to loads above its maximum rating, the material of the load cell may plastically deform; this may result in a signal offset, loss of linearity, difficulty with or impossibility of calibration, or even mechanical damage to the sensing element (e.g. delamination, rupture).

- Wiring issues: the wires to the cell may develop high resistance, e.g. due to corrosion. Alternatively, parallel current paths can be formed by ingress of moisture. In both cases the signal develops offset (unless all wires are affected equally) and accuracy is lost.

- Electrical damage: the load cells can be damaged by induced or conducted current. Lightnings hitting the construction, or arc welding performed near the cells, can overstress the fine resistors of the strain gauges and cause their damage or destruction. For welding nearby, it is suggested to disconnect the load cell and short all its pins to the ground, nearby the cell itself. High voltages can break through the insulation between the substrate and the strain gauges.

- Nonlinearity: at the low end of their scale, the load cells tend to be nonlinear. This becomes important for cells sensing very large ranges, or with large surplus of load capability to withstand temporary overloads or shocks (e.g. the rope clamps). More points may be needed for the calibration curve.

Excitation and Rated Output

The bridge is excited with stabilized voltage (usually 10V, but can be 20V, 5V, or less for battery powered instrumentation). The difference voltage proportional to the load then appears on the signal outputs. The cell output is rated in millivolts per volt (mV/V) of the difference voltage at full rated mechanical load. So a 2.96 mV/V load cell will provide 29.6 millivolt signal at full load when excited with 10 volts.

Typical sensitivity values are 1 to 3 mV/V. Typical maximum excitation voltage is around 15 volts.

Wiring

The full-bridge cells come typically in four-wire configuration. The wires to the top and bottom end of the bridge are the excitation (often labelled E+ and E−, or Ex+ and Ex−), the wires to its sides are the signal (labelled S+ and S−). Ideally, the voltage difference between S+ and S− is zero under zero load, and grows proportionally to the load cell's mechanical load.

Sometimes a six-wire configuration is used. The two additional wires are "sense" (Sen+ and Sen−), and are connected to the bridge with the Ex+ and Ex- wires, in a fashion similar to four-terminal sensing. With these additional signals, the controller can compensate for the change in wire resistance due to e.g. temperature fluctuations.

The individual resistors on the bridge usually have resistance of 350 Ω. Sometimes other values (typically 120 Ω, 1,000 Ω) can be encountered.

The bridge is typically electrically insulated from the substrate. The sensing elements are in close proximity and in good mutual thermal contact, to avoid differential signals caused by temperature differences.

Using Multiple Cells

One or more load cells can be used for sensing a single load.

If the force can be concentrated to a single point (small scale sensing, ropes, tensile loads, point loads), a single cell can be used. For long beams, two cells at the end are used. Vertical cylinders can be measured at three points, rectangular objects usually require four sensors. More sensors are used for large containers or platforms, or very high loads.

If the loads are guaranteed to be symmetrical, some of the load cells can be substituted with pivots. This saves the cost of the load cell but can significantly decrease accuracy.

Load cells can be connected in parallel; in that case, all the corresponding signals are connected together (Ex+ to Ex+, S+ to S+, ...), and the resulting signal is the average of the signals from all the sensing elements. This is often used in e.g. personal scales, or other multipoint weight sensors.

Wiring Colors

The most common color assignment is red for Ex+, black for Ex−, green for S+, and white for S−.

Less common assignments are red for Ex+, white for Ex−, green for S+, and blue for S−, or red for Ex+, blue for Ex−, green for S+, and yellow for S−. Other values are also possible, e.g. red for Ex+, green for Ex−, yellow for S+ and blue for S−.

Piezoelectric Load Cell

Piezoelectric load cells work on the same principle of deformation as the strain gauge load cells, but a voltage output is generated by the basic piezoelectric material - proportional to the deformation of load cell. Useful for dynamic/frequent measurements of force. Most applications for piezo-based load cells are in the dynamic loading conditions, where strain gauge load cells can fail with high dynamic loading cycles.

Hydraulic Load Cell

The cell uses conventional piston and cylinder arrangement. The piston is placed in a thin elastic diaphragm. The piston doesn't actually come in contact with the load cell. Mechanical stops are placed to prevent over strain of the diaphragm when the loads exceed certain limit. The load cell is completely filled with oil. When the load is applied on the piston, the movement of the piston and the diaphragm results in an increase of oil pressure which in turn produces a change in the pressure on a Bourdon tube connected with the load cells. Because this sensor has no electrical components, it is ideal for use in hazardous areas. Typical hydraulic load cell applications include tank, bin, and hopper weighing. By example, a hydraulic load cell is immune to transient voltages (lightning) so these type of load cells might be a more effective device in outdoor environments. This technology is more expensive than other types of load cells. It is a more costly technology and thus cannot effectively compete on a cost of purchase basis.

Pneumatic Load Cell

The load cell is designed to automatically regulate the balancing pressure. Air pressure is applied to one end of the diaphragm and it escapes through the nozzle placed at the bottom of the load cell. A pressure gauge is attached with the load cell to measure the pressure inside the cell. The deflection of the diaphragm affects the airflow through the nozzle as well as the pressure inside the chamber.

Other Types

Other types include vibrating wire load cells, which are useful in geomechanical applications due to low amounts of drift, and capacitive load cells where the capacitance of a capacitor changes as the load presses the two plates of a capacitor closer together.

Ringing

Every load cell is subject to "ringing" when subjected to abrupt load changes. This stems from the spring-like behavior of load cells. In order to measure the loads, they have to deform. As such, a load cell of finite stiffness must have spring-like behavior, exhibiting vibrations at its natural frequency. An oscillating data pattern can be the result of ringing. Ringing can be suppressed in a limited fashion by passive means. Alternatively, a control system can use an actuator to actively damp out the ringing of a load cell. This method offers better performance at a cost of significant increase in complexity.

Uses

Load cells are used in several types of measuring instruments such as laboratory balances, industrial scales, platform scales and universal testing machines. From 1993 the British Antarctic Survey installed load cells in glass fibre nests to weigh albatross chicks. Load cells are used in a wide variety of items such as the seven-post shaker which is often used to set up race cars.

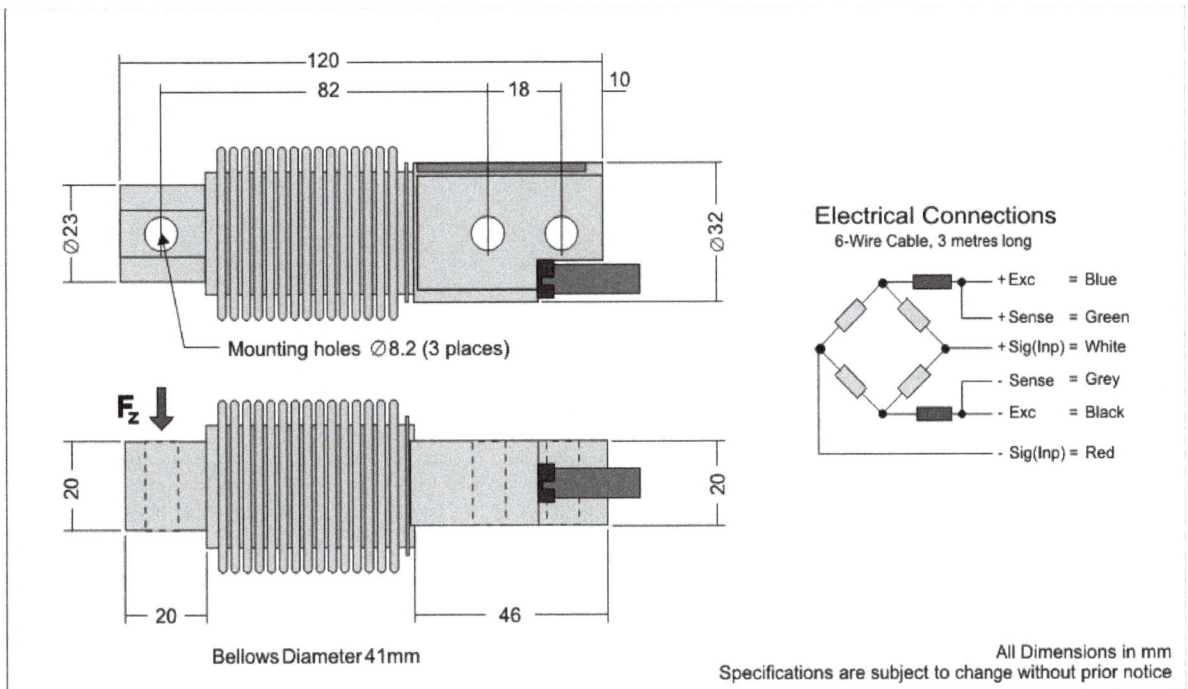

Mounting holes ∅8.2 (3 places)

F_z

Bellows Diameter 41mm

Electrical Connections
6-Wire Cable, 3 metres long

+ Exc = Blue
+ Sense = Green
+ Sig(Inp) = White
- Sense = Grey
- Exc = Black
- Sig(Inp) = Red

All Dimensions in mm
Specifications are subject to change without prior notice

DS355-4, 12/09

References

- O'Leary, Beth Laura; Darrin, Ann Garrison (2009). Handbook of Space Engineering, Archaeology, and Heritage. Hoboken: CRC Press. pp. 239–240. ISBN 9781420084320.

- Michael Russell Rip; James M. Hasik (2002). The Precision Revolution: GPS and the Future of Aerial Warfare. Naval Institute Press. p. 65. ISBN 1-55750-973-5. Retrieved January 14, 2010.

- Misra, Pratap; Enge, Per (2006). Global Positioning System. Signals, Measurements and Performance (2nd ed.). Ganga-Jamuna Press. p. 115. ISBN 0-9709544-1-7. Retrieved August 16, 2013.

- Borre, Kai; M. Akos, Dennis; Bertelsen, Nicolaj; Rinder, Peter; Jensen, Søren Holdt (2007). A Software-Defined GPS and Galileo Receiver. A single-Frequency Approach. Springer. p. 18. ISBN 0-8176-4390-7.

- Dargie, W. and Poellabauer, C. (2010). Fundamentals of wireless sensor networks: theory and practice. John Wiley and Sons. pp. 168–183, 191–192. ISBN 978-0-470-99765-9.

- Sohraby, K., Minoli, D., Znati, T. (2007). Wireless sensor networks: technology, protocols, and applications. John Wiley and Sons. pp. 203–209. ISBN 978-0-471-74300-2.

- Seitz, Frederick (1999). The cosmic inventor: Reginald Aubrey Fessenden (1866-1932). 89. American Philosophical Society. pp. 41–46. ISBN 0-87169-896-X.

- Turner, Anthony; Wilson, George; Kaube, Isao (1987). Biosensors:Fundamentals and Applications. Oxford, UK: Oxford University Press. p. 770. ISBN 0198547242.

- Bănică, Florinel-Gabriel (2012). Chemical Sensors and Biosensors:Fundamentals and Applications. Chichester, UK: John Wiley & Sons. p. 576. ISBN 9781118354230.

- Grosvenor, Edwin S. and Wesson, Morgan. Alexander Graham Bell: The Life and Times of the Man Who Invented the Telephone. New York: Harry N. Abrahms, Inc., 1997. ISBN 0-8109-4005-1.

Varied Types of Actuators

Actuators can best be understood in confluence with the major topics listed in the following chapter. Actuators are of five types, hydraulic, pneumatic, electoral, thermal and mechanical actuators. Actuators in machines are responsible for moving or controlling the system. The aspects elucidated in this chapter are of vital importance, and provide a better understanding of actuators.

Pneumatic Actuator

A pneumatic actuator converts energy (typically in the form of compressed air) into mechanical motion. The motion can be rotary or linear, depending on the type of actuator. Some types of pneumatic actuators include:

- Tie rod cylinders

- Rotary actuators

- Grippers

- Rodless actuators with magnetic linkage or rotary cylinders

- Rodless actuators with mechanical linkage

- Pneumatic artificial muscles

- Speciality actuators that combine rotary and linear motion—frequently used for clamping operations

- Vacuum generators

A Pneumatic actuator mainly consists of a piston, a cylinder, and valves or ports. The piston is covered by a diaphragm, or seal, which keeps the air in the upper portion of the cylinder, allowing air pressure to force the diaphragm downward, moving the piston underneath, which in turn moves the valve stem, which is linked to the internal parts of the actuator. Pneumatic actuators may only have one spot for a signal input, top or bottom, depending on action required. Valves require little pressure to operate and usually double or *triple* the input force. The larger the size of the piston, the larger the output pressure can be. Having a larger piston can also be good if air supply is low, allowing the same forces with less input. These pressures are large enough to crush objects in the pipe. On 100 kPa input, you could lift a small car (upwards of 1,000 lbs) easily, and this is only a basic, small pneumatic valve. However, the resulting forces required of the stem would be too great and cause the valve stem to fail.

This pressure is transferred to the valve stem, which is hooked up to either the valve plug, butterfly valve etc. Larger forces are required in high pressure or high flow pipelines to allow the valve to overcome these forces, and allow it to move the valves moving parts to control the material flowing inside.

Valves input pressure is the "control signal." This can come from a variety of measuring devices, and each different pressure is a different set point for a valve. A typical standard signal is 20–100 kPa. For example, a valve could be controlling the pressure in a vessel which has a constant out-flow, and a varied in-flow (varied by the actuator and valve). A pressure transmitter will monitor the pressure in the vessel and transmit a signal from 20–100 kPa. 20 kPa means there is no pressure, 100 kPa means there is full range pressure (can be varied by the transmiters calibration points). As the pressure rises in the vessel, the output of the transmitter rises, this increase in pressure is sent to the valve, which causes the valve to stroke downward, and start closing the valve, decreasing flow into the vessel, reducing the pressure in the vessel as excess pressure is evacuated through the out flow. This is called a direct acting process.

Types of Pneumatic Actuator

Pneumatic Gripper

A pneumatic gripper is a specific type of pneumatic actuator that typically involves either parallel or angular motion of surfaces, A.K.A. "tooling jaws or fingers" that will grip an object. When combined with other pneumatic, electric, or hydraulic components, the gripper can be used as part of a "pick and place" system that will allow a component to be picked up and placed somewhere else as part of a manufacturing system.

Some grippers act directly on the object they are gripping based on the force of the air pressure supplied to the gripper, while others will use a mechanism such as a gear or toggle to leverage the amount of force applied to the object being gripped. Grippers can also vary in terms of the opening size, the amount of force that can be applied, and the shape of the gripping surfaces—frequently called "tooling jaws or fingers". They can be used to pick up everything from very small items (a transistor or chip for a circuit board, for example) to very large items, such as an engine block for a car. Grippers are frequently added to industrial robots in order to allow the robot to interact with other objects.

Common industrial pneumatic components include:

- pneumatic direct operated solenoid valve
- pneumatic pilot operated solenoid valve
- pneumatic external piloted solenoid valve
- pneumatic manual valve
- pneumatic valve with air pilot actuator
- pneumatic filter
- pneumatic pressure regulator
- pneumatic lubricator

- pneumatic pressure switch

- pneumatic manual OSHA-type lock out and dump valve

- pneumatic solenoid dump valve

- pneumatic rodless cylinder

- pneumatic gripper

- pneumatic rotary actuator

- pneumatic fitting

- pneumatic flow control

- pneumatic quick exhaust valve

- pneumatic pressure booster

- pneumatic polyurethane tubing

- pneumatic quick disconnect

Hydraulic Cylinder

A hydraulic cylinder (also called a linear hydraulic motor) is a mechanical actuator that is used to give a unidirectional force through a unidirectional stroke. It has many applications, notably in construction equipment (engineering vehicles), manufacturing machinery, and civil engineering.

The hydraulic cylinders on this excavator operate the machine's linkages.

Operation

Hydraulic cylinders get their power from pressurized hydraulic fluid, which is typically oil. The hydraulic cylinder consists of a cylinder barrel, in which a piston connected to a piston rod moves back and forth. The barrel is closed on one end by the cylinder bottom (also called the cap) and the other end by the cylinder head (also called the gland) where the piston rod comes out of the cylinder. The piston has sliding rings and seals. The piston divides the inside of the cylinder into two chambers, the bottom chamber (cap end) and the piston rod side chamber (rod end / head end).

Flanges, trunnions, clevises, Lugs are common cylinder mounting options. The piston rod also has mounting attachments to connect the cylinder to the object or machine component that it is pushing / pulling.

A hydraulic cylinder is the actuator or "motor" side of this system. The "generator" side of the hydraulic system is the hydraulic pump which brings in a fixed or regulated flow of oil to the hydraulic cylinder, to move the piston. The piston pushes the oil in the other chamber back to the reservoir. If we assume that the oil enters from cap end, during extension stroke, and the oil pressure in the rod end / head end is approximately zero, the force F on the piston rod equals the pressure P in the cylinder times the piston area A:

$$F = P \cdot A$$

Retraction Force Difference

For double-acting single-rod cylinders, when the input and output pressures are reversed, there is a force difference between the two sides of the piston due to one side of the piston being covered by the rod attached to it. The cylinder rod reduces the surface area of the piston and reduces the force that can be applied for the retraction stroke.

During the retraction stroke, if oil is pumped into the head (or gland) at the rod end and the oil from the cap end flows back to the reservoir without pressure, the fluid pressure in the rod end is (Pull Force) / (piston area - piston rod area):

$$P = \frac{F_p}{A_p - A_r}$$

where P is the fluid pressure, F_p is the pulling force, A_p is the piston face area and A_r is the rod cross-section area.

For double-acting, double-rod cylinders, when the piston surface area is equally covered by a rod of equal size on both sides of the head, there is no force difference. Such cylinders typically have their cylinder body affixed to a stationary mount.

Parts of A Hydraulic Cylinder

A hydraulic cylinder consists of the following parts:

Cylinder Barrel

The main function of cylinder body is to hold cylinder pressure. The cylinder barrel is mostly made from a seamless tube. The cylinder barrel is ground and/or honed internally with a typical surface finish of 4 to 16 microinch. Normally hoop stress is calculated to optimize the barrel size. The piston reciprocates in the cylinder.

Cylinder Base or Cap

The main function of the cap is to enclose the pressure chamber at one end. The cap is connected to the body by means of welding, threading, bolts, or tie rod. Caps also perform as cylinder mounting components [cap flange, cap trunnion, cap clevis]. Cap size is determined based on the bending stress. A static seal / o-ring is used in between cap and barrel (except welded construction).

Cylinder Head

The main function of the head is to enclose the pressure chamber from the other end. The head contains an integrated rod sealing arrangement or the option to accept a seal gland. The head is connected to the body by means of threading, bolts, or tie rod. A static seal / o-ring is used in between head and barrel.

Piston

The main function of the piston is to separate the pressure zones inside the barrel. The piston is machined with grooves to fit elastomeric or metal seals and bearing elements. These seals can be single acting or double acting. The difference in pressure between the two sides of the piston causes the cylinder to extend and retract. The piston is attached with the piston rod by means of threads, bolts, or nuts to transfer the linear motion.

Piston Rod

The piston rod is typically a hard chrome-plated piece of cold-rolled steel which attaches to the piston and extends from the cylinder through the rod-end head. In double rod-end cylinders, the actuator has a rod extending from both sides of the piston and out both ends of the barrel. The piston rod connects the hydraulic actuator to the machine component doing the work. This connection can be in the form of a machine thread or a mounting attachment.

Seal Gland

The cylinder head is fitted with seals to prevent the pressurized oil from leaking past the interface between the rod and the head. This area is called the seal gland. The advantage of a seal gland is easy removal and seal replacement. The seal gland contains a primary seal, a secondary seal / buffer seal, bearing elements, wiper / scraper and static seal. In some cases, especially in small hydraulic cylinders, the rod gland and the bearing elements are made from a single integral machined part.

Seals

The seals are considered / designed as per the cylinder working pressure, cylinder speed, operating temperature, working medium and application. Piston seals are dynamic seals, and they can be single acting or double acting. Generally speaking, Elastomer seals made from nitrile rubber, Polyurethane or other materials are best in lower temperature environments, while seals made of Fluorocarbon Viton are better for higher temperatures. Metallic seals are also available and commonly use cast iron for the seal material. Rod seals are dynamic seals and generally are single acting. The compounds of rod seals are nitrile rubber, Polyurethane, or Fluorocarbon Viton. Wipers / scrapers are used to eliminate contaminants such as moisture, dirt, and dust, which can cause extensive damage to cylinder walls, rods, seals and other components. The common compound for wipers is polyurethane. Metallic scrapers are used for sub zero temperature applications, and applications where foreign materials can deposit on the rod. The bearing elements / wear bands are used to eliminate metal to metal contact. The wear bands are designed as per the side load requirements. The primary compounds for wear bands are filled PTFE, woven fabric reinforced polyester resin and bronze

Other Parts

There are many component parts that make up the internal portion of a hydraulic cylinder. All of these pieces combine to create a fully functioning component.

- Cylinder base connection

- Cushions

- Internal Threaded Ductile Heads

- Head Glands

- Polypak Pistons

- Cylinder Head Caps

- Butt Plates

- Eye Brackets/Clevis Brackets

- MP Detachable Mounts

- Rod Eyes/Rod Clevis

- Pivot Pins

- Spherical Ball Bushings

- Spherical Rod Eye

- Alignment Coupler

- Ports and Fittings

Single Acting Vs. Double Acting

- Single acting cylinders are economical and the simplest design. Hydraulic fluid enters through a port at one end of the cylinder, which extends the rod by means of area difference. An external force or gravity returns the piston rod.

- Double acting cylinders have a port at each end or side of the piston, supplied with hydraulic fluid for both the retraction and extension.

Hydraulic Cylinder Designs

There are primarily two styles of hydraulic cylinder construction used in industry: Tie rod style cylinders and welded body style cylinders.

Tie Rod Cylinder

Hydraulic Cylinder

Tie rod style hydraulic cylinders use high strength threaded steel rods to hold the two end caps to the cylinder barrel. This method of construction is most often seen in industrial factory applications. Small bore cylinders usually have 4 tie rods, while large bore cylinders may require as many as 16 or 20 tie rods in order to retain the end caps under the tremendous forces produced. Tie rod style cylinders can be completely disassembled for service and repair, and they are not always customizable.

A tie rod cylinder

The National Fluid Power Association (NFPA) has standardized the dimensions of hydraulic tie rod cylinders. This enables cylinders from different manufacturers to interchange within the same mountings.

Welded Body Cylinder

Welded body cylinders have no tie rods. The barrel is welded directly to the end caps. The ports are welded to the barrel. The front rod gland is usually threaded into or bolted to the cylinder barrel. This allows the piston rod assembly and the rod seals to be removed for service.

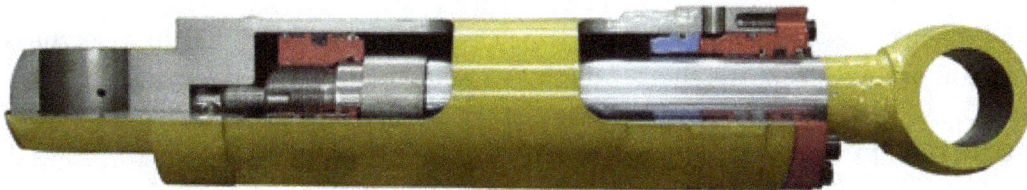

A Cut Away of a Welded Body Hydraulic Cylinder showing the internal components

Welded body cylinders have a number of advantages over tie rod style cylinders. Welded cylinders have a narrower body and often a shorter overall length enabling them to fit better into the tight confines of machinery. Welded cylinders do not suffer from failure due to tie rod stretch at high pressures and long strokes. The welded design also lends itself to customization. Special features are easily added to the cylinder body, including special ports, custom mounts, valve manifolds, and so on.

The smooth outer body of welded cylinders also enables the design of multi-stage telescopic cylinders.

Welded body hydraulic cylinders dominate the mobile hydraulic equipment market such as construction equipment (excavators, bulldozers, and road graders) and material handling equipment (forklift trucks, telehandlers, and lift-gates). They are also used in heavy industry such as cranes, oil rigs, and large off-road vehicles in above-ground mining.

Piston Rod Construction

The piston rod of a hydraulic cylinder operates both inside and outside the barrel, and consequently both in and out of the hydraulic fluid and surrounding atmosphere.

Coatings

Wear and corrosion resistant surfaces are desirable on the outer diameter of the piston rod. The surfaces are often applied using coating techniques such as Chrome (Nickel) Plating, Lunac 2+ duplex, Laser Cladding, PTA welding and Thermal Spraying. These coatings can be finished to the desirable surface roughness (Ra, Rz) where the seals show optimum performance. All these coating methods have their specific advantages and disadvantages. It is for this reason that coating experts play a crucial role in selecting the optimum surface treatment procedure for protecting Hydraulic Cylinders.

Cylinders are used in different operational conditions and that makes it a challenge to find the right coating solution. In dredging there might be impact from stones or other parts, in salt water environments there are extreme corrosion attacks, in off-shore cylinders facing bending and impact in combination with salt water, and in the steel industry there are high temperatures involved, etc.... It is important to understand that currently there is no single coating solution which successfully combats all the specific operational wear conditions. Every single technique has its own benefits and disadvantages.

Length

Piston rods are generally available in lengths which are cut to suit the application. As the common rods have a soft or mild steel core, their ends can be welded or machined for a screw thread.

Distribution of Forces on Components

The forces on the piston face and the Piston Head Retainer vary depending on what Piston Head retention system is used.

If a circlip (or any non preloaded system) is used, the force acting to separate the Piston Head and the Cylinder Shaft shoulder is the applied pressure multiplied by the area of the Piston Head.

The Piston Head and Shaft shoulder will separate and the load is fully reacted by the Piston Head Retainer.

If a preloaded system is used the force between the Cylinder Shaft and Piston Head is initially the Piston Head Retainer preload value. Once pressure is applied this force will reduce. The Piston Head and Cylinder Shaft shoulder will remain in contact unless the applied pressure multiplied by Piston Head area exceeds the preload.

The maximum force the Piston Head Retainer will see is the larger of the preload and the applied pressure multiplied by the full Piston Head area. It is interesting to note that the load on the Piston Head Retainer is greater than the external load, which is due to the reduced shaft size passing through the Piston Head. Increasing this portion of shaft reduces the load on the Retainer.

Side Loading

Side loading is unequal pressure that is not centered on the cylinder rod. This off-center strain can lead to bending of the rod in extreme cases, but more commonly causes leaking due to warping the circular seals into an oval shape. It can also damage and enlarge the bore hole around the rod and the inner cylinder wall around the piston head, if the rod is pressed hard enough sideways to fully compress and deform the seals to make metal-on-metal scraping contact.

The strain of side loading can be directly reduced with the use of internal stop tubes which reduce the maximum extension length, leaving some distance between the piston and bore seal, and increasing leverage to resist warping of the seals. Double pistons also spread out the forces of side loading while also reducing stroke length. Alternately, external sliding guides and hinges can support the load and reduce side loading forces applied directly on the cylinder.

Repair

Hydraulic cylinders form the most integral part of many hydraulic systems. It is a common practice to dissemble and rebuild an entire device in the case of hydraulic cylinder repair. Inspection of the leakage issue and scrutinizing cylinder parts (especially the seals) is helpful in recognizing the exact problem and choosing on the repair options accordingly. Steps involved in the repair of hydraulic cylinders:

Disassembly

First of all, you should place the cylinder in a suitable location, which has sufficient space to work. If you are working in a cumbersome space, it will be difficult for you to keep track of opened up parts. After bringing the cylinder to an appropriate spot, open the cylinder ports and drain out all the hydraulic fluid. The cover of cylinder can be removed by unscrewing the bolts. Once you take off the cover, remove the piston by loosening the input valves.

Diagnosis

Once the piston is completely removed, you will be able to see multiple seals on different parts that are connected to the piston rod . First of all, you need to examine the piston rod to see if there is any damage. If the shaft of the rod has bends or if the Cylinder Bore has scratches, then get them repaired at a professional repair shop. If the damage is permanent, then you can order and man-ufacture new piston rod for your hydraulic cylinder. Sometimes, piston seals get greatly damaged, distorted, or broken. Such damaged seals can cause leakage of hydraulic fluid from the cylinder leading and lower the overall pressure. When such events occur, you know that these seals need to be replaced.

Repairing or Replacing Damaged Parts

The parts of the hydraulic cylinder that are distorted (piston rod, rod seal, piston seal and/ or head of rod), need to be either repaired or completely replaced with new parts. The seals can be re-packed with the help of a hydraulic cylinder seal kit. These kits will have seals and suitable o-rings. Remember the size and type of old seal while removing them and fix the new ones accordingly. Make sure that you handle the new seals with utmost care so that they do not get damaged in any way.

Rebuilding

Before reassembling all the parts of your cylinder, you should clean and dry the cylinder barrel completely. Also clean the piston rod, shaft, and other parts of the cylinder. Get the broken and damaged seals repacked. Then assemble the parts back on the piston rod. The assembly needs to be done in a reverse order. Once you have assembled all the parts back, put the rod into the soft-jaw vise and screw back the bolts onto the piston rod.

Important Tip

If the parts of the hydraulic cylinder are severely damaged, then it is advisable to replace them with new parts with the help of a professional repair expert. Trying to replace/ repair too many parts on your own can lead to faulty reassembly. By following the above steps, you can accomplish the task of hydraulic cylinder repair. Make sure that you prevent ingress of moisture or dirt after assembly of the parts is done.

Cylinder Mounting Methods

Mounting methods also play an important role in cylinder performance. Generally, fixed mounts on the centerline of the cylinder are best for straight line force transfer and avoiding wear. Common types of mounting include:

Flange mounts—Very strong and rigid, but have little tolerance for misalignment. Experts rec-ommend cap end mounts for thrust loads and rod end mounts where major loading puts the piston rod in tension. Three types are head rectangular flange, head square flange or rectan-gular head. Flange mounts function optimally when the mounting face attaches to a machine support member.

Side-mounted cylinders—Easy to install and service, but the mounts produce a turning moment as the cylinder applies force to a load, increasing wear and tear. To avoid this, specify a stroke at least as long as the bore size for side mount cylinders (heavy loading tends to make short stroke, large bore cylinders unstable). Side mounts need to be well aligned and the load supported and guided.

Centerline lug mounts —Absorb forces on the centerline, and require dowel pins to secure the lugs to prevent movement at higher pressures or under shock conditions. Dowel pins hold it to the machine when operating at high pressure or under shock loading.

Pivot mounts —Absorb force on the cylinder centerline and let the cylinder change alignment in one plane. Common types include clevises, trunnion mounts and spherical bearings. Because these mounts allow a cylinder to pivot, they should be used with rod-end attachments that also pivot. Clevis mounts can be used in any orientation and are generally recommended for short strokes and small- to medium-bore cylinders.

Special Hydraulic Cylinders

Telescopic Cylinder

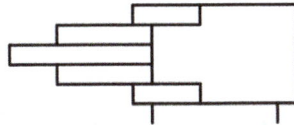

Telescopic cylinder (ISO 1219 symbol)

The length of a hydraulic cylinder is the total of the stroke, the thickness of the piston, the thickness of bottom and head and the length of the connections. Often this length does not fit in the machine. In that case the piston rod is also used as a piston barrel and a second piston rod is used. These kinds of cylinders are called telescopic cylinders. If we call a normal rod cylinder single stage, telescopic cylinders are multi-stage units of two, three, four, five or more stages. In general telescopic cylinders are much more expensive than normal cylinders. Most telescopic cylinders are single acting (push). Double acting telescopic cylinders must be specially designed and manufactured.

Plunger Cylinder

A hydraulic cylinder without a piston or with a piston without seals is called a plunger cylinder. A plunger cylinder can only be used as a pushing cylinder; the maximum force is piston rod area multiplied by pressure. This means that a plunger cylinder in general has a relatively thick piston rod.

Differential Cylinder

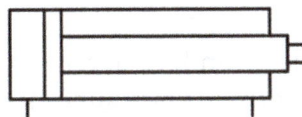

Differential cylinder (ISO 1219 symbol)

A differential cylinder acts like a normal cylinder when pulling. If the cylinder however has to push, the oil from the piston rod side of the cylinder is not returned to the reservoir, but goes to the bottom side of the cylinder. In such a way, the cylinder goes much faster, but the maximum force the cylinder can give is like a plunger cylinder. A differential cylinder can be manufactured like a normal cylinder, and only a special control is added.

Position Sensing "Smart" Hydraulic Cylinder

Position sensing hydraulic cylinders eliminate the need for a hollow cylinder rod. Instead, an external sensing "bar" using Hall Effect technology senses the position of the cylinder's piston. This is accomplished by the placement of a permanent magnet within the piston. The magnet propagates a magnetic field through the steel wall of the cylinder, providing a locating signal to the sensor.

Terminology

In the United States, popular usage refers to the whole assembly of cylinder, piston, and piston rod (or more) collectively as a "piston", which is incorrect. Instead, the piston is the short, cylindrical metal component that separates the two parts of the cylinder barrel internally.

Part and Hydraulic Cylinder

Hydraulic Motor

A hydraulic motor is a mechanical actuator that converts hydraulic pressure and flow into torque and angular displacement (rotation). The hydraulic motor is the rotary counterpart of the hydraulic cylinder.

A small hydraulic motor

Conceptually, a hydraulic motor should be interchangeable with a hydraulic pump because it performs the opposite function - similar to the way a DC electric motor is theoretically interchangeable with a DC electrical generator. However, most hydraulic pumps cannot be used as hydraulic motors because they cannot be backdriven. Also, a hydraulic motor is usually designed for working pressure at both sides of the motor.

Hydraulic pumps, motors, and cylinders can be combined into hydraulic drive systems. One or more hydraulic pumps, coupled to one or more hydraulic motors, constitute a hydraulic transmission.

One of the first rotary hydraulic motors to be developed was that constructed by William Armstrong for his Swing Bridge over the River Tyne. Two motors were provided, for reliability. Each one was a three-cylinder single-acting oscillating engine. Armstrong developed a wide range of hydraulic motors, linear and rotary, that were used for a wide range of industrial and civil engineering tasks, particularly for docks and moving bridges.

The first simple fixed-stroke hydraulic motors had the disadvantage that they used the same volume of water whatever the load and so were wasteful at part-power. Unlike steam engines, as water is incompressible, they could not be throttled or their valve cut-off controlled. To overcome this, motors with variable stroke were developed. Adjusting the stroke, rather than controlling admission valves, now controlled the engine power and water consumption. One of the first of these was Arthur Rigg's patent engine of 1886. This used a double eccentric mechanism, as used on variable stroke power presses, to control the stroke length of a three cylinder radial engine. Later, the swashplate engine with an adjustable swashplate angle would become a popular way to make variable stroke hydraulic motors.

Swing Bridge, River Tyne

Hydraulic Motor Types

Many designs are possible. The following types of hydraulic motors are available:

Symbol hydraulic motor

Gear and Vane Motors

Gear and vane motors are used in simple rotating systems. Their benefits include low initial cost and high rpm.

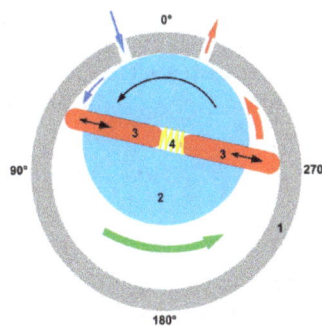

Gear motor

A gear motor (external gear) consists of two gears, the driven gear (attached to the output shaft by way of a key, etc.) and the idler gear. High pressure oil is ported into one side of the gears, where it flows around the periphery of the gears, between the gear tips and the wall housings in which it resides, to the outlet port. The gears then mesh, not allowing the oil from the outlet side to flow back to the inlet side. For lubrication, the gear motor uses a small amount of oil from the pressurized side of the gears, bleeds this through the (typically) hydrodynamic bearings, and vents the same oil either to the low pressure side of the gears, or through a dedicated drain port on the motor housing. An especially positive attribute of the gear motor is that catastrophic breakdown is less common than in most other types of hydraulic motors. This is because the gears gradually wear down the housing and/or main bushings, reducing the volumetric efficiency of the motor gradually until it is all but useless. This often happens long before wear causes the unit to seize or break down.

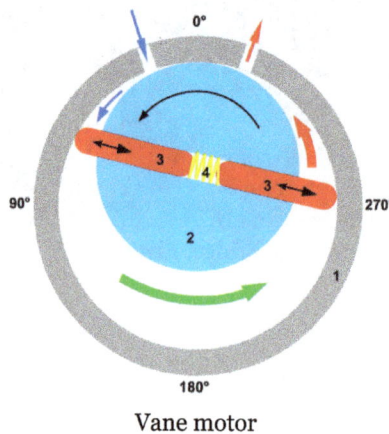

Vane motor

A vane motor consists of a housing with an eccentric bore, in which runs a rotor with vanes in it that slide in and out. The force differential created by the unbalanced force of the pressurized fluid on the vanes causes the rotor to spin in one direction. A critical element in vane motor design is how the vane tips are machined at the contact point between vane tip and motor housing. Several types of "lip" designs are used, and the main objective is to provide a tight seal between the inside of the motor housing and the vane, and at the same time to minimize wear and metal-to-metal contact.v

Gerotor Motors

The gerotor motor is in essence a rotor with N-1 teeth, rotating off center in a rotor/stator with N teeth. Pressurized fluid is guided into the assembly using a (usually) axially placed plate-type distributor valve. Several different designs exist, such as the Geroller (internal or external rollers) and Nichols motors. Typically, the Gerotor motors are low-to-medium speed and medium-to-high torque.

Axial Plunger Motors

For high quality rotating drive systems plunger motors are generally used. Whereas the speed of hydraulic pumps range from 1200 to 1800 rpm, the machinery to be driven by the motor often requires a much lower speed. This means that when an axial plunger motor (swept volume maximum 2 litres) is used, a gearbox is usually needed. For a continuously adjustable swept volume, axial piston motors are used. PISTON TYPE.— Like piston (reciprocating) type pumps, the most common design of the piston type of motor is the axial. This type of motor is the most commonly used in hydraulic systems. These motors are, like their pump counterparts, available in both variable and fixed displacement designs. Typical usable (within acceptable efficiency) rotational speeds range from below 50 rpm to above 14000 rpm. Efficiencies and minimum/maximum rotational speeds are highly dependent on the design of the rotating group, and many different types are in use.

Radial Piston Motors

Staffa hydraulic motor

Radial piston motors are available in two basic types, pistons pushing inward, and pistons pushing outward.

Calzoni hydraulic motor

The crankshaft type (e.g. Staffa or Sai hydraulic motors) with a single cam and the pistons pushing inwards is basically an old design but is one which has extremely high starting torque characteristics. They are available in displacements from 40 cc/rev up to about 50 litres/rev but can sometimes be limited in power output. Crankshaft type radial piston motors are capable of running at "creep" speeds and some can run seamlessly up to 1500 rpm whilst offering virtually constant output torque characteristics. This makes them still the most versatile design.

The single-cam-type radial piston motor exists in many different designs itself. Usually the difference lies in the way the fluid is distributed to the different pistons or cylinders, and also the design of the cylinders themselves. Some motors have pistons attached to the cam using rods (much like in an internal combustion engine), while others employ floating "shoes", and even spherical contact telescopic cylinders like the Parker Denison Calzoni type. Each design has its own set of pros and cons, such as freewheeling ability, high volumetric efficiency, high reliability and so on.

Multi-lobe cam ring types (e.g. Rexroth, Hägglunds Drives, Poclain, or Eaton Hydre-MAC type) have a cam ring with multiple lobes and the piston rollers push *outward* against the cam ring. This produces a very smooth output with high starting torque but they are often limited in the upper speed range. This type of motor is available in a very wide range from about 1 litre/rev to 250 litres/rev. These motors are particularly good on low speed applications and can develop very high power.

Braking

Hydraulic motors usually have a drain connection for the internal leakage, which means that when the power unit is turned off the hydraulic motor in the drive system will move slowly if an external load is acting on it. Thus, for applications such as a crane or winch with suspended load, there is always a need for a brake or a locking device.

Uses

Hydraulic motors are used for many applications now such as winches and crane drives, wheel motors for military vehicles, self-driven cranes, and excavators. Conveyor and feeder drives, mixer and agitator drives, roll mills, drum drives for digesters, trommels and kilns, shredders for cars, tyres, cable and general garbage, drilling rigs, trench cutters, high-powered lawn trimmers, plastic injection machines

Linear Actuator

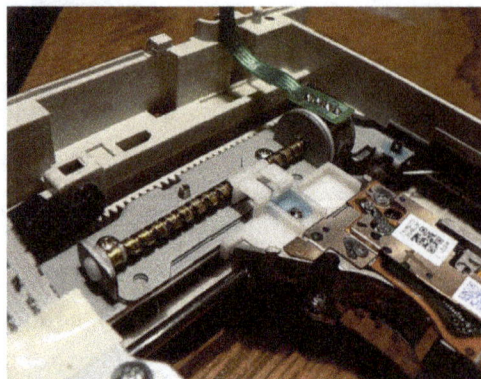

DVD drive with leadscrew and stepper motor.

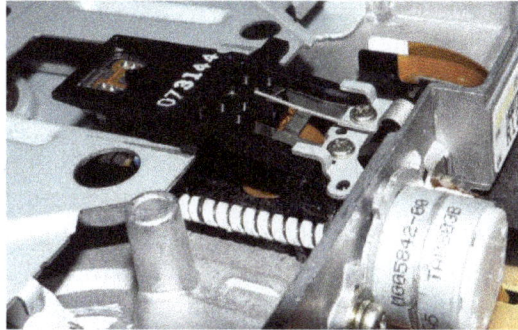

Floppy disc drive with leadscrew and stepper motor.

A linear actuator is an actuator that creates motion in a straight line, in contrast to the circular motion of a conventional electric motor. Linear actuators are used in machine tools and industrial machinery, in computer peripherals such as disk drives and printers, in valves and dampers, and in many other places where linear motion is required. Hydraulic or pneumatic cylinders inherently produce linear motion. Many other mechanisms are used to generate linear motion from a rotating motor.

Types

Mechanical Actuators

Mechanical linear actuators typically operate by conversion of rotary motion into linear motion. Conversion is commonly made via a few simple types of mechanism:

A mechanical linear actuator with digital readout (a type of micrometer).

- Screw: leadscrew, screw jack, ball screw and roller screw actuators all operate on the principle of the simple machine known as the screw. By rotating the actuator's nut, the screw shaft moves in a line.

- Wheel and axle: Hoist, winch, rack and pinion, chain drive, belt drive, rigid chain and rigid belt actuators operate on the principle of the wheel and axle. A rotating wheel moves a cable, rack, chain or belt to produce linear motion.

- Cam: Cam actuators function on a principle similar to that of the wedge, but provide relatively limited travel. As a wheel-like cam rotates, its eccentric shape provides thrust at the base of a shaft.

Some mechanical linear actuators only pull, such as hoists, chain drive and belt drives. Others only push (such as a cam actuator). Pneumatic and hydraulic cylinders, or lead screws can be designed to generate force in both directions.

Mechanical actuators typically convert rotary motion of a control knob or handle into linear displacement via screws and/or gears to which the knob or handle is attached. A jackscrew or car jack is a familiar mechanical actuator. Another family of actuators are based on the segmented spindle. Rotation of the jack handle is converted mechanically into the linear motion of the jack head. Mechanical actuators are also frequently used in the field of lasers and optics to manipulate the position of linear stages, rotary stages, mirror mounts, goniometers and other positioning instruments. For accurate and repeatable positioning, index marks may be used on control knobs. Some actuators include an encoder and digital position readout. These are similar to the adjustment knobs used on micrometers except their purpose is position adjustment rather than position measurement.

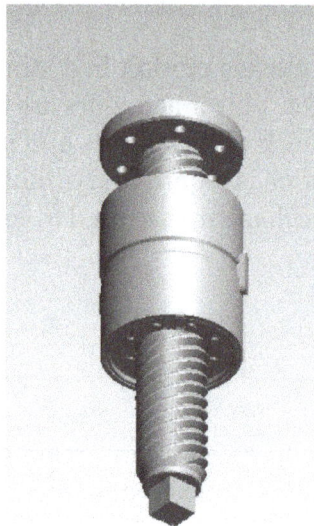

Roller screw actuation with traveling screw (rotating nut).

Hydraulic Actuators

Hydraulic actuators or hydraulic cylinders typically involve a hollow cylinder having a piston inserted in it. An unbalanced pressure applied to the piston generates force that can move an external object. Since liquids are nearly incompressible, a hydraulic cylinder can provide controlled precise linear displacement of the piston. The displacement is only along the axis of the piston. A familiar example of a manually operated hydraulic actuator is a hydraulic car jack. Typically though, the term "hydraulic actuator" refers to a device controlled by a hydraulic pump.

Pneumatic Actuators

Pneumatic actuators, or pneumatic cylinders, are similar to hydraulic actuators except they use compressed gas to generate force instead of a liquid. They work similarly to a piston in which air is pumped inside a chamber and pushed out of the other side of the chamber. Air actuators are not necessarily used for heavy duty machinery and instances where large amounts of weight are present. One of the reasons pneumatic linear actuators are preferred to other types is the fact that the power source is simply an air compressor. Because air is the input source, pneumatic actuators are able to be used in many places of mechanical activity. The downside is, most air compressors are large, bulky, and loud. They are hard to transport to other areas once installed. Pneumatic linear actuators are likely to leak and this makes them less efficient than mechanical linear actuators.

Piezoelectric Actuators

The piezoelectric effect is a property of certain materials in which application of a voltage to the material causes it to expand. Very high voltages correspond to only tiny expansions. As a result, piezoelectric actuators can achieve extremely fine positioning resolution, but also have a very short range of motion. In addition, piezoelectric materials exhibit hysteresis which makes it difficult to control their expansion in a repeatable manner.

Electro-Mechanical Actuators

A miniature electro-mechanical linear actuator where the lead nut is part of the motor. The lead screw does not rotate, so as the lead nut is rotated by the motor, the lead screw is extended or retracted.

Pressure-compensated underwater linear actuator, used on a Remotely Operated Underwater Vehicle (ROV)

Typical compact cylindrical linear electric actuator

Typical linear or rotary + linear electric actuator

Electro-mechanical actuators are similar to mechanical actuators except that the control knob or handle is replaced with an electric motor. Rotary motion of the motor is converted to linear displacement. There are many designs of modern linear actuators and every company that manufactures them tends to have a proprietary method. The following is a generalized description of a very simple electro-mechanical linear actuator.

Simplified Design

Typically, an electric motor is mechanically connected to rotate a lead screw. A lead screw has a continuous helical thread machined on its circumference running along the length (similar to the thread on a bolt). Threaded onto the lead screw is a lead nut or ball nut with corresponding helical threads. The nut is prevented from rotating with the lead screw (typically the nut interlocks with a non-rotating part of the actuator body). Therefore, when the lead screw is rotated, the nut will be driven along the threads. The direction of motion of the nut depends on the direction of rotation of the lead screw. By connecting linkages to the nut, the motion can be converted to usable linear displacement. Most current actuators are built for high speed, high force, or a compromise between the two. When considering an actuator for a particular application, the most important specifications are typically travel, speed, force, accuracy, and lifetime. Most varieties are mounted on dampers or butterfly valves.

There are many types of motors that can be used in a linear actuator system. These include dc brush, dc brushless, stepper, or in some cases, even induction motors. It all depends on the application requirements and the loads the actuator is designed to move. For example, a linear actuator using an integral horsepower AC induction motor driving a lead screw can be used to operate a large valve in a refinery. In this case, accuracy and high movement resolution aren't needed, but high force and speed are. For electromechanical linear actuators used in laboratory instrumentation robotics, optical and laser equipment, or X-Y tables, fine resolution in the micron range and high accuracy may require the use of a fractional horsepower stepper motor linear actuator with a fine pitch lead screw. There are many variations in the electromechanical linear actuator system. It is critical to understand the design requirements and application constraints to know which one would be best.

Standard Vs Compact Construction

A linear actuator using standard motors will commonly have the motor as a separate cylinder attached to the side of the actuator, either parallel with the actuator or perpendicular to the actuator. The motor may be attached to the end of the actuator. The drive motor is of typical construction with a solid drive shaft that is geared to the drive nut or drive screw of the actuator.

Compact linear actuators use specially designed motors that try to fit the motor and actuator into the smallest possible shape.

- The inner diameter of the motor shaft can be enlarged, so that the drive shaft can be hollow. The drive screw and nut can therefore occupy the center of the motor, with no need for additional gearing between the motor and the drive screw.

- Similarly the motor can be made to have a very small outside diameter, but instead the pole faces are stretched lengthwise so the motor can still have very high torque while fitting in a small diameter space.

Principles

In the majority of linear actuator designs, the basic principle of operation is that of an inclined plane. The threads of a lead screw act as a continuous ramp that allows a small rotational force to be used over a long distance to accomplish movement of a large load over a short distance.

Variations

Many variations on the basic design have been created. Most focus on providing general improvements such as a higher mechanical efficiency, speed, or load capacity. There is also a large engineering movement towards actuator miniaturization.

Most electro-mechanical designs incorporate a lead screw and lead nut. Some use a ball screw and ball nut. In either case the screw may be connected to a motor or manual control knob either directly or through a series of gears. Gears are typically used to allow a smaller (and weaker) motor spinning at a higher rpm to be geared down to provide the torque necessary to spin the screw under a heavier load than the motor would otherwise be capable of driving directly. Effectively this sacrifices actuator speed in favor of increased actuator thrust. In some applications the use of worm gear is common as this allow a smaller built in dimension still allowing great travel length.

A traveling-nut linear actuator has a motor that stays attached to one end of the lead screw (perhaps indirectly through a gear box), the motor spins the lead screw, and the lead nut is restrained from spinning so it travels up and down the lead screw.

A traveling-screw linear actuator has a lead screw that passes entirely through the motor. In a traveling-screw linear actuator, the motor "crawls" up and down a lead screw that is restrained from spinning. The only spinning parts are inside the motor, and may not be visible from the outside.

Some lead screws have multiple "starts". This means they have multiple threads alternating on the same shaft. One way of visualizing this is in comparison to the multiple color stripes on a candy cane. This allows for more adjustment between thread pitch and nut/screw thread contact area, which determines the extension speed and load carrying capacity (of the threads), respectively.

Static Load Capacity

Linear screw actuators can have a static loading capacity, meaning that when the motor stops the actuator essentially locks in place and can support a load that is either pulling or pushing on the actuator. This static load capacity increases mobility and speed.

The braking force of the actuator varies with the angular pitch of the screw threads and the specific design of the threads. Acme threads have a very high static load capacity, while ball screws have an extremely low load capacity and can be nearly free-floating.

Generally it is not possible to vary the static load capacity of screw actuators without additional technology. The screw thread pitch and drive nut design defines a specific load capacity that cannot be dynamically adjusted.

In some cases, high viscosity grease can be added to linear screw actuators to increase the static load. Some manufacturers use this to alter the load for specific needs.

Static load capacity can be added to a linear screw actuator using an electromagnetic brake system, which applies friction to the spinning drive nut. For example, a spring may be used to apply brake pads to the drive nut, holding it in position when power is turned off. When the actuator needs to be moved, an electromagnet counteracts the spring and releases the braking force on the drive nut.

Similarly an electromagnetic ratchet mechanism can be used with a linear screw actuator so that the drive system lifting a load will lock in position when power to the actuator is turned off. To lower the actuator, an electromagnet is used to counteract the spring force and unlock the ratchet.

Dynamic Load Capacity

Dynamic load capacity is typically referred to as the amount of force the linear actuator is capable of providing during operation. This force will vary with screw type (amount of friction restricting movement) and the motor driving the movement. Dynamic load is the figure which most actuators are classified by, and is a good indication of what applications it would suit best.

Speed Control

In most cases when using an electro-mechanical actuator, it is preferred to have some type of speed control. Such controllers vary the voltage supplied to the motor, which in turn changes the speed at which the lead screw turns.

Duty Cycle

The duty cycle of a motor refers to the amount of time the actuator can be run before it needs to cool down. Staying within this guideline when operating an actuator is key to its longevity and performance. If the duty cycle rating is exceeded, then overheating, loss of power, and eventual burning of the motor is risked.

Linear Motors

A linear motor is functionally the same as a rotary electric motor with the rotor and stator circular magnetic field components laid out in a straight line. Where a rotary motor would spin around and re-use the same magnetic pole faces again, the magnetic field structures of a linear motor are physically repeated across the length of the actuator.

Since the motor moves in a linear fashion, no lead screw is needed to convert rotary motion to linear. While high capacity is possible, the material and/or motor limitations on most designs are surpassed relatively quickly due to a reliance solely on magnetic attraction and repulsion forces. Most linear motors have a low load capacity compared to other types of linear actuators. Linear motors have an advantage in outdoor or dirty environments in that the two halves do not need to contact each other, and so the electromagnetic drive coils can be waterproofed and sealed against moisture and corrosion, allowing for a very long service life.

Telescoping Linear Actuator

Telescoping linear actuators are specialized linear actuators used where space restrictions exist. Their range of motion is many times greater than the unextended length of the actuating member.

Rigid chain actuator

A common form is made of concentric tubes of approximately equal length that extend and retract like sleeves, one inside the other, such as the telescopic cylinder.

Other more specialized telescoping actuators use actuating members that act as rigid linear shafts when extended, but break that line by folding, separating into pieces and/or uncoiling when retracted. Examples of telescoping linear actuators include:

- Helical band actuator

- Rigid belt actuator

- Rigid chain actuator

- Segmented spindle

Advantages and Disadvantages

Actuator Type	Advantages	Disadvantages
Mechanical	Cheap. Repeatable. No power source required. Self-contained. Identical behaviour extending or retracting.	Manual operation only. No automation.
Electro-mechanical	Cheap. Repeatable. Operation can be automated. Self-contained. Identical behaviour extending or retracting. DC or stepping motors. Position feedback possible.	Many moving parts prone to wear.
Linear motor	Simple design. Minimum of moving parts. High speeds possible. Self-contained. Identical behaviour extending or retracting.	Low to medium force.
Piezoelectric	Very small motions possible at high speeds. Consumes barely any power.	Short travel unless amplified mechanically. High voltages required, typically 24V or more. Expensive, and fragile. Good in compression only, not in tension. Typically used for Fuel Injectors
Hydraulic	Very high forces possible.	Can leak. Requires position feedback for repeatability. External hydraulic pump required. Some designs good in compression only.
Pneumatic	Strong, light, simple, fast.	Precise position control impossible except at full stops

Wax motor	Smooth operation.	Not as reliable as other methods.
Segmented spindle	Very compact. Range of motion greater than length of actuator.	Both linear and rotary motion.
Moving coil	Force, position and speed are controllable and repeatable. Capable of high speeds and precise positioning. Linear, rotary, and linear + rotary actions possible.	Requires position feedback to be repeatable.
MICA (moving iron controllable actuator)	High force and controllable. Higher force and less losses than moving coils. Losses easy to dissipate. Electronic driver easy to design and set up.	Stroke limited to several millimeters, less linearity than moving coils

Types of Linear Actuator

Ball Screw

A ball screw is a mechanical linear actuator that translates rotational motion to linear motion with little friction. A threaded shaft provides a helical raceway for ball bearings which act as a precision screw. As well as being able to apply or withstand high thrust loads, they can do so with minimum internal friction. They are made to close tolerances and are therefore suitable for use in situations in which high precision is necessary. The ball assembly acts as the nut while the threaded shaft is the screw. In contrast to conventional leadscrews, ballscrews tend to be rather bulky, due to the need to have a mechanism to re-circulate the balls.

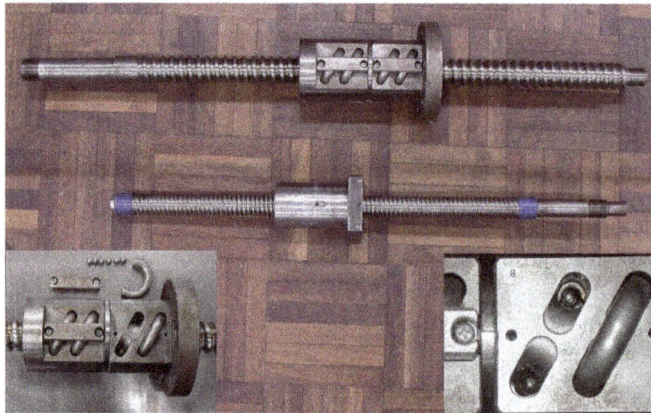

Photo showing two ball screws. Inset images are close-up photos of the ball assembly of the top screw. Left inset: recirculating tube removed showing retainer bracket, loose balls and tube. Right inset: closer view of the nut cavity.

Another form of linear actuator based on a rotating rod is the threadless ballscrew, a.k.a. "rolling ring drive". In this design, three (or more) rolling-ring bearings are arranged symmetrically in a housing surrounding a smooth (thread-less) actuator rod or shaft. The bearings are set at an angle to the rod, and this angle determines the direction and rate of linear motion per revolution of the rod. An advantage of this design over the conventional ballscrew or leadscrew is the practical elimination of backlash and loading caused by preload nuts.

Applications

Ball screws are used in aircraft and missiles to move control surfaces, especially for electric fly by wire, and in automobile power steering to translate rotary motion from an electric motor to axial motion of the steering rack. They are also used in machine tools, robots and precision assembly equipment. High precision ball screws are used in steppers for semiconductor manufacturing.

History

Historically, the first precise screwshafts were produced by starting with a low precision screwshaft, and then lapping the shaft with several spring-loaded nut laps. By rearranging and inverting the nut laps, the lengthwise errors of the nuts and shaft were averaged. Then, the very repeatable shaft's pitch is measured against a distance standard. A similar process is sometimes used today to produce reference standard screw shafts, or master manufacturing screw shafts.

Description and Operation

To maintain their inherent accuracy and ensure long life, great care is needed to avoid contamination with dirt and abrasive particles. This may be achieved by using rubber or leather bellows to completely or partially enclose the working surfaces. Another solution is to use a positive pressure of filtered air when they are used in a semi-sealed or open enclosure.

While reducing friction, ball screws can operate with some preload, effectively eliminating backlash (slop) between input (rotation) and output (Linear motion). This feature is essential when they are used in computer-controlled motion-control systems, e.g., CNC machine tools and high precision motion applications (e.g., wire bonding).

Disadvantages

Depending upon their lead angle, ball screws can be back-driven due to their low internal friction (i.e., the screw shaft can be driven linearly to rotate the ball nut). They are usually undesirable for hand-fed machine tools, as the stiffness of a servo motor is required to keep the cutter from grabbing the work and self-feeding, that is, where the cutter and workpiece exceed the optimum feedrate and effectively jam or crash together, ruining the cutter and workpiece. Cost is also a major factor as Acme screws are cheaper to manufacture.

Advantages

Low friction in ball screws yields high mechanical efficiency compared to alternatives. A typical ball screw may be 90 percent efficient, versus 20 to 25 percent efficiency of an Acme lead screw of equal size. Lack of sliding friction between the nut and screw lends itself to extended lifespan of the screw assembly (especially in no-backlash systems), reducing downtime for maintenance and parts replacement, while also decreasing demand for lubrication. This, combined with their overall performance benefits and reduced power requirements, may offset the initial costs of using ball screws.

Manufacture

Ball screw shafts may be fabricated by rolling, yielding a less precise, but inexpensive and mechanically efficient product. Rolled ball screws have a positional precision of several thousandths of an inch per foot.

Accuracy

High-precision screw shafts are typically precise to one thousandth of an inch per foot (830 nanometers per centimeter) or better. They have historically been machined to gross shape, case hardened and then ground. The three step process is needed because high temperature machining distorts the work-piece. Hard whirling is a recent (2008) precision machining technique that minimizes heating of the work, and can produce precision screws from case-hardened bar stock.

Instrument quality screw shafts are typically precise to 250 nanometers per centimeter. They are produced on precision milling machines with optical distance measuring equipment and special tooling. Similar machines are used to produce optical lenses and mirrors. Instrument screw shafts are generally made of Invar, to prevent temperature from changing tolerances too much.

Ball Return Systems

The circulating balls travel inside the screw and nut thread. If the ball nut did not have a return mechanism the balls would fall out of the end of the ball nut when they reached the end of the nut. For this reason several different recirculation methods have been developed.

An external ballnut employs a stamped tube which picks up balls from raceway with use of small pick up finger. Balls travel inside of tube and are then replaced back in thread raceway.

An internal button ballnut employs a machined or cast button style return which allows balls to exit raceway track and move one thread and reenter raceway.

An endcap return ballnut employs a cap on the end of ball nut. The cap is machined to pick up balls out of the end of nut and direct them down holes which are bored transversely down the ballnut. The compliment cap on the other side of nut directs balls back into raceway.

Thread Profile

To obtain proper rolling action of the balls, as in a standard ball bearing, it is necessary that, when loaded in one direction, the ball makes contact at one point with the nut, and one point with the screw. In practice, most ball screws are designed to be lightly preloaded, so that there is at least a slight load on the ball at four points, two in contact with the nut and two in contact with the screw. This is accomplished by using a thread profile that has a slightly larger radius than the ball, the difference in radii being kept small (e.g. a simple V thread with flat faces is unsuitable) so that elastic deformation around the point of contact allows a small, but non-zero contact area to be obtained, like any other rolling element bearing. To this end, the threads are usually machined as a "gothic arch" profile. If a simple semicircular thread profile were used, contact would only be at two points, on the outer and inner edges, which would not resist axial loading.

Preloading

To remove backlash and obtain the optimum stiffness and wear characteristics for a given application, a controlled amount of preload is usually applied. This is accomplished in some cases by machining the components such that the balls are a "tight" fit when assembled, however this gives poor control of the preload, and can not be adjusted to allow for wear. It is more common to design the ball nut as effectively two separate nuts which are tightly coupled mechanically, with adjustment by either rotation one nut with respect to the other, so creating a relative axial displacement, or by retaining both nuts tightly together axially and rotating one with respect to the other, so that its set of balls is displaced axially to create the preload.

Equations

$$T = \frac{Fl}{2\pi v}$$

Where T is torque applied to screw or nut, F is linear force applied, l is ball screw lead, and v is ball screw efficiency.

Ball Screw Standards

National and international standards are used to standardize the definitions, environmental requirements, and test methods used for ball screws. Selection of the standard to be used is an agreement between the supplier and the user and has some significance in the design of the screw. In the United States, ASME has developed the B5.48-1977 Standard entitled "Ball Screws".

Rack and Pinion

A rack and pinion is a type of linear actuator that comprises a pair of gears which convert rotational motion into linear motion. A circular gear called "the pinion" engages teeth on a linear "gear" bar called "the rack"; rotational motion applied to the pinion causes the rack to move relative to the pinion, thereby translating the rotational motion of the pinion into linear motion.

Rack and pinion animation

For example, in a rack railway, the rotation of a pinion mounted on a locomotive or a railcar engages a rack between the rails and forces a train up a steep slope.

For every pair of conjugate involute profile, there is a basic rack. This basic rack is the profile of the conjugate gear of infinite pitch radius. (I.e. a toothed straight edge.)

A generating rack is a rack outline used to indicate tooth details and dimensions for the design of a generating tool, such as a hob or a gear shaper cutter.

Applications

Rack and pinion combinations are often used as part of a simple linear actuator, where the rotation of a shaft powered by hand or by a motor is converted to linear motion.

Lock gate controls on a canal

The rack carries the full load of the actuator directly and so the driving pinion is usually small, so that the gear ratio reduces the torque required. This force, thus torque, may still be substantial and so it is common for there to be a reduction gear immediately before this by either a gear or worm gear reduction. Rack gears have a higher ratio, thus require a greater driving torque, than screw actuators.

Stairlifts

Most Stairlifts today are operating using the Rack & Pinion system.

Steering

A rack and pinion is commonly found in the steering mechanism of cars or other wheeled, steered vehicles. Rack and pinion provides a less efficient mechanical advantage than other mechanisms such as recirculating ball, but less backlash and greater feedback, or steering "feel". The mechanism may be power-assisted, usually by hydraulic or electrical means.

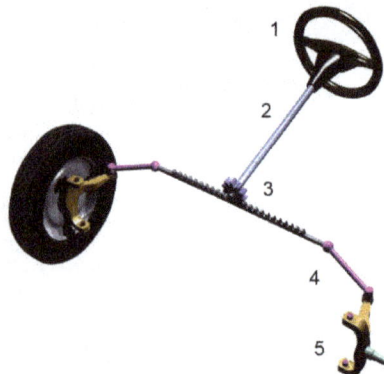

Rack steering in an automobile

The use of a variable rack (still using a normal pinion) was invented by Arthur Ernest Bishop, in the 1970s, so as to improve vehicle response and steering "feel," especially at high speeds. He also created a low cost press forging process to manufacture the racks, eliminating the need to machine the gear teeth.

Rack Railways

Rack railways are mountain railways that use a rack built into the centre of the track and a pinion on their locomotives. This allows them to work on steep gradients, up to 1 in 2 (50%), far in excess of those a conventional railway relying on friction alone can achieve.

Rack railway axle

Although the extra grip of the rack system is obviously important for climbing, it has perhaps a more important use in also allowing controlled braking on these steep lines and for being much less affected by snow or ice on the rails.

Actuators

A rack and pinion with two racks and one pinion is used in actuators. An example is pneumatic rack and pinion actuators that can be used to control valves in pipeline transport. The actuators in the picture on the right are used to control the valves of large water pipeline. In the top actuator, a gray control signal line can be seen connecting to a solenoid valve (the small black box attached to the back of the top actuator), which is used as the pilot for the actuator. The solenoid valve controls the air pressure coming from the input air line (the small green tube). The output air from the solenoid valve is fed to the chamber in the middle of the actuator, increasing the pressure. The pressure in the actuator's chamber pushes the pistons away. While the pistons are moving apart from each other, the attached racks are also moved along the pistons in the opposite directions of the two racks. The two racks are meshed to a pinion at the direct opposite teeth of the pinion. When the two racks move, the pinion is turned, causing the attached main valve of the water pipe to turn.

Comb Drive

Comb-drives are actuators, often used as ganern linear actuatorselectrostatic forces that act between two electrically conductive combs. Comb drive actuators typically operate at the micro- or nanometer scale and are generally manufactured by bulk micromachining or surface micromachining a silicon wafer substrate.

The attractive electrostatic forces are created when a voltage is applied between the static and moving combs causing them to be drawn together. The force developed by the actuator is proportional to the change in capacitance between the two combs, increasing with driving voltage, the number of comb teeth, and the gap between the teeth. The combs are arranged so that they never touch (because then there would be no voltage difference). Typically the teeth are arranged so that they can slide past one another until each tooth occupies the slot in the opposite comb.

Restoring springs, levers, and crankshafts can be added if the motor's linear operation is to be converted to rotation or other motions.

The force can be derived by first starting with the energy stored in a capacitor and then differentiating in the direction of the force. The energy in a capacitor is given by:

$$E = \frac{1}{2}CV^2$$

$$F = \frac{1}{2}\frac{\partial C}{\partial x_{drive}}V^2$$

Using the capacitance for a parallel plate capacitor, the force is:

$$F = \frac{-1}{2}\frac{nt\epsilon_o\epsilon_r V^2}{d^2}$$

V = applied electric potential, $\dot{o}_r$ = relative permittivity of dielectric, ϵ_o = permittivity of free space (8.85 pF/m), n = total number of fingers on both sides of electrodes, t = thickness in the out of plane direction of the electrodes, d = gap between electrodes.

Structure of Comb-drives

• rows of interlocking teeth • half fixed • half part of movable assembly • electrically isolated • electrostatic attraction/ repulsion – CMOS drive voltage • many teeth increased force – typically 10μm long and strong

Scaling Issues

Comb drives cannot scale to large gap distances (equivalently actuation distance), since development of effective forces at large gaps distances would require high voltages—therefore limited by electrical breakdown. More importantly, limitations imposed by gap distance limits the actuation distance.

Rigid Chain Actuator

A rigid chain actuator, known variously as a linear chain actuator, push-pull chain actuator, electric chain actuator or column-forming chain actuator, is a specialized mechanical linear actuator used in window operating, push-pull material handling and lift applications. The actuator is a chain and pinion device that forms an articulated telescoping member to transmit traction and thrust. High-capacity rigid chain lifting columns (jacks) can move dynamic loads exceeding 10 tonnes (US 20,000 pounds) over more than 7 metres (20 feet) of travel.

Principle of Operation

Rigid chain actuators function as rack and pinion linear actuators that use articulated racks. Rigid chain actuators use limited-articulation chains, usually resembling a roller chain, that engage with pinions mounted on a drive shaft within a housing. The links of the actuating member, the "rigid chain", are articulated in a manner that they deflect from a straight line to one side only. As the pinions spin, the links of the chain are rotated 90 degrees through the housing, which guides and locks the chain into a rigid linear form effective at resisting tension and compression (buckling). Because the actuating member can fold on itself, it can be stored relatively compactly in a storage magazine, either in an overlapping or coiled arrangement. Rigid chain actuators are generally driven by electric motors. Most rigid chains are manufactured from steel.

U.S. Patent   May 1, 2001   Sheet 1 of 2   US 6,224,037 B1

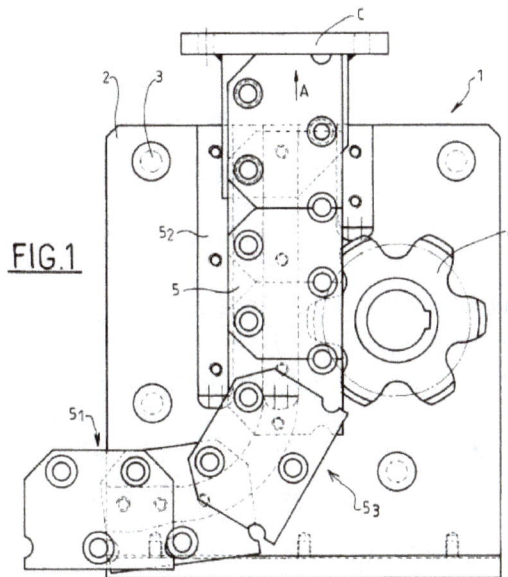

FIG.1

Patent Drawing for Serapid LinkLift Rigid Chain Actuator (2001).

Use

Modified roller chain has been used extensively in material handling equipment, but could only be used in push-pull applications when a continuous loop of chain was used (with the exception of chain encapsulated in a guide channel). The development of efficient rigid chain actuators broadened the use of chain actuation for industrial applications. Small scale rigid chain actuators are used as building hardware, incorporated into windows, door and hatches as motorized open/close mechanisms. Rigid chain actuators are also used as the lifting columns in performing arts facilities, incorporated in stage, orchestra and seating platform lift systems.

Increasingly rigid chain systems are being incorporated into scissor lift tables or platform lifts as the method of actuation, replacing hydraulic cylinders. They are also used for production line automation and die changing.

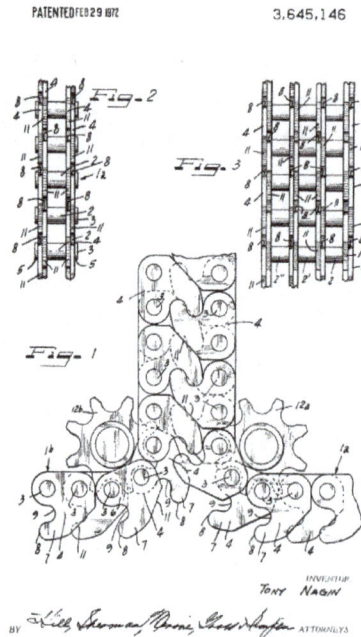

Patent Drawing for Interlocking Rigid Chain Actuator (1972).

Types

The primary distinction between types of rigid chain actuator is whether the actuating member is formed from a single chain or from a pair of interlocking chains in a back-to-back arrangement, like a zipper. Interlocking chain actuators have the advantages over single-chain actuators of improved resistance to buckling and that the actuating member does not require lateral restraint at its leading end in order to resist a modicum of transverse loads on any edge of the member. For example, it may function as a relatively stable telescoping pole.

The design of the chain varies significantly depending on application and manufacturer. Variants have been designed to, among other things:

- Simplify manufacture

- Reduce friction and maintenance

- Limit size and weight

- Increase speed, travel, capacity, efficiency and stability

Development

Rigid chain actuators were developed from "chain rammers" that used a single "ram chain" thrust from a magazine to load heavy-caliber ordnance into the breech of a cannon. Robert Matthews received a US patent for his "Mechanical Rammer" in 1901 which used a roller on the leading end of the chain to guide it and allow thrust without deflection. Developed more than a century ago, his rammer still bears a strong resemblance to many modern rigid chain actuators. In 1908 Oscar Knoch was awarded a US patent for his "Chain Rammer for Guns". By orienting the folding side of the chain upward his ram chain acted as a self-supporting telescoping beam with negligible sag. Used in this manner the need for a separate guide was eliminated.

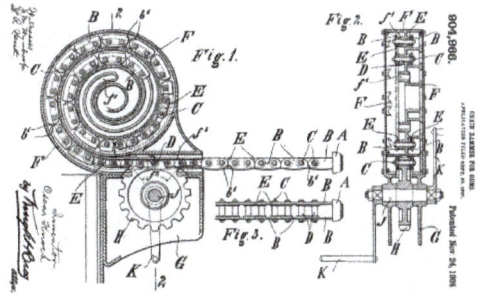

Patent Drawing for Chain Rammer (1908).

An early conception of chain used as a telescoping column instead a horizontal rammer was by Eldridge E. Long, who was awarded a US patent for his "Lifting Jack" in 1933, which he believed was "particularly adapted for use upon automobiles". It used a double chain configuration, each chain linking solid bearing blocks that were stacked to resist compressive loads. In 1951, Yaichi Hayakawa was awarded a US patent for his "Interlocking Chain Stanchion" which eliminated bearing blocks by integrating the compressive path of force into the interlocking links of two roller-like chains. The zipper action of back-to-back interlocking chains provided guideless chain travel regardless of orientation and path of travel.

It should be noted that in 1941, prior to the double chain configuration, Karl Bender received a US patent for "Compression Resistant Chain" using three interlocking chains. In addition to the back-to-back arrangement of the typical interlocking chain actuator, a third chain was interlocked between the other two at a right angle. Perhaps due to their relative complexity, triple-chain actuators are not common.

Rigid Belt Actuator

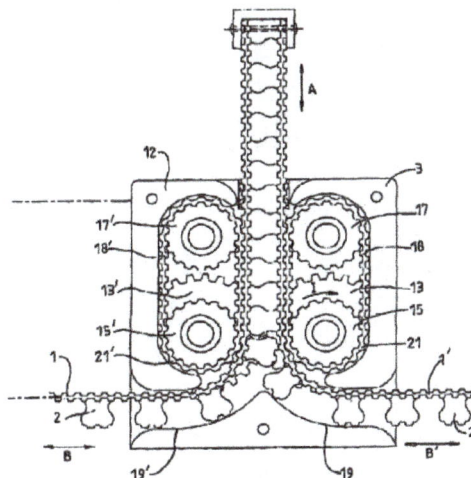

FIG.1

Patent Drawing for Serapid RigiBelt Rigid Belt Actuator (2007).

A rigid belt actuator, also known as a push-pull belt actuator or zipper belt actuator, is a specialized mechanical linear actuator used in push-pull and lift applications. The actuator is a belt and pinion device that forms a telescoping beam or column member to transmit traction and thrust. Rigid belt actuators can move dynamic loads up to approximately 230 pounds over about 3 feet of travel.

Principle of Operation

Rigid belt actuators can be thought of as rack and pinion devices that use a flexible rack. Rigid belt actuators use two reinforced plastic ribbed belts, that engage with pinions mounted on drive shafts within a housing. The belts have evenly spaced load bearing blocks on the non-ribbed face. As the pinions spin, the belts are rotated 90 degrees through the housing, which interlocks the blocks like a zipper into a rigid linear form. The resulting beam or column is effective at resisting tension and compression (buckling). Because the actuating member can fold on itself, it can be stored relatively compactly in a storage magazine, either in an overlapping or coiled arrangement. The actuator is driven by an electric motor.

Development

A rigid belt actuator is effectively a non-metallic variation of the rigid chain actuator. But, while the interlocking chain actuator has been around since the middle of the 20th century, rigid belt technology didn't emerge until the new millennium. Joël Bourc'His received a patent for his "Linear Belt Actuator" in 2007.

Helical Band Actuator

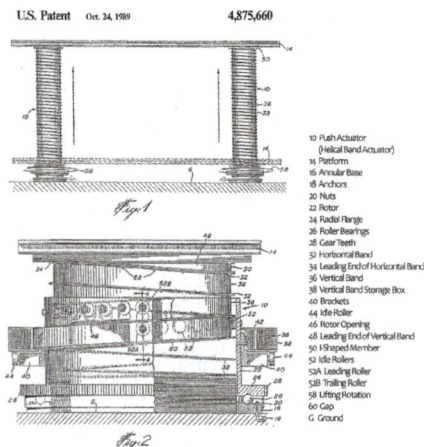

U.S. Patent   Oct. 24, 1989   4,875,660

10 Push Actuator
   (Helical Band Actuator)
14 Platform
16 Annular Base
18 Anchors
20 Nuts
22 Rotor
24 Radial Flange
26 Roller Bearings
28 Gear Teeth
32 Horizontal Band
34 Leading End of Horizontal Band
36 Vertical Band
38 Vertical Band Storage Box
40 Brackets
44 Idle Roller
46 Rotor Opening
48 Leading End of Vertical Band
50 I-Shaped Member
52 Idle Rollers
52A Leading Roller
52B Trailing Roller
58 Lifting Rotation
60 Gap
G Ground

Patent Drawing for Spiralift Actuator(1989), with legend.

A helical band actuator, generally known by the trademark Spiralift, is a complex and specialized linear actuator used in stage lifts and material handling lifts. The actuator forms a high-capacity telescoping tubular column (lifting capacities to 25,000 pounds, travel to 40 feet).

Raison D'être

Pierre Gagnon was awarded a US patent for the "Push Actuator" in 1989. Gagnon developed the actuator to substitute hydraulic cylinders in stage and orchestra lift systems. Hydraulic jacks were the predominant push actuators used in performing arts facilities, but they had their issues.

The piston rods of a hydraulic lift extend below the platform they support by more than the length that the lift travels. For lifts at grade, housings for the rods are often sunk into support holes, known as caissons, that are inaccessible and often complicated by groundwater levels. If placed above ground, the space occupied by the rod housings is unusable. Gagnon's actuator eliminates caissons and reduces the volume of unusable space below the platform of the lift.

Another significant issue with hydraulic lifts is platform drift, the tendency of a platform to lower due to depressurization of the hydraulic system over time. Helical band actuators do not suffer from platform drift because they are rigid mechanical devices.

Traveling nut screw jacks and rigid chain actuators can also be effective at addressing these issues with hydraulic jacks, and are also used for stage lifts.

Principle of Operation

The telescoping column is formed by a pair of interlocking stainless steel bands. One band has a vertical rectangular profile and the other horizontal, much like an oversized Slinky. The vertical band spirals up on itself into a stacked helix, forming the wall of the column, while at the same time, the horizontal band interlocks the continuous spiral seam of the vertical band. When the column lowers, the bands separate and retract into two compact coils. The bands are combined, separated and stored by an assembly located at the base of the column. The result is an efficient (50%-80%) telescoping lifting column.

To incorporate the device into a lift system, multiple helical band actuators are arranged below the lift platform where they are powered by an electric motor(s) and synchronized transmission. Helical band actuators require a separate lateral support mechanism, usually provided in the form of guide rails or self-guiding frame, such as a pantograph (e.g. a scissor lift).

Plasma Actuator

Plasma actuators are a type of actuator currently being developed for aerodynamic flow control. Plasma actuators impart force in a similar way to ionocraft.

The working of these actuators is based on the formation of a low-temperature plasma between a pair of asymmetric electrodes by application of a high-voltage AC signal across the electrodes. Consequently, air molecules from the air surrounding the electrodes are ionized, and are accelerated through the electric field.

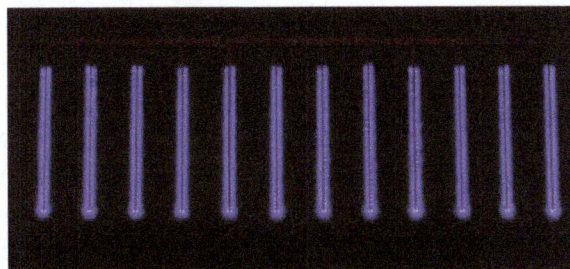

Glow of plasma actuator discharges

Introduction

Plasma actuators operating at the atmospheric conditions are promising for flow control, mainly for their physical properties, such as the induced body force by a strong electric field and the generation of heat during an electric arc, and the simplicity of their constructions and placements. In particular, the recent invention of glow discharge plasma actuators by Roth (2003) that can produce sufficient quantities of glow discharge plasma in the atmosphere pressure air helps to yield an increase in flow control performance.

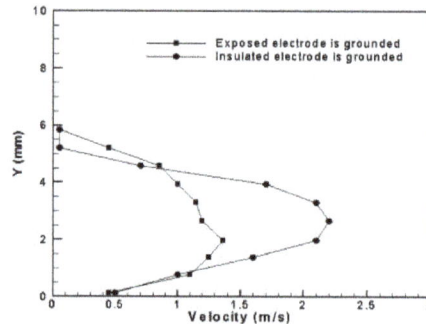

Local flow speed induced by a plasma actuator

Power Supply and Electrode Layouts

Either a direct current (DC) or an alternating current(AC) power supply or a microwave microdischarge can be used for different configurations of plasma actuators. One schematic of an AC power supply design for a dielectric barrier discharge plasma actuator is given here as an example. The performance of plasma actuators is determined by dielectric materials and power inputs, later is limited by the qualities of MOSFET or IGBT.

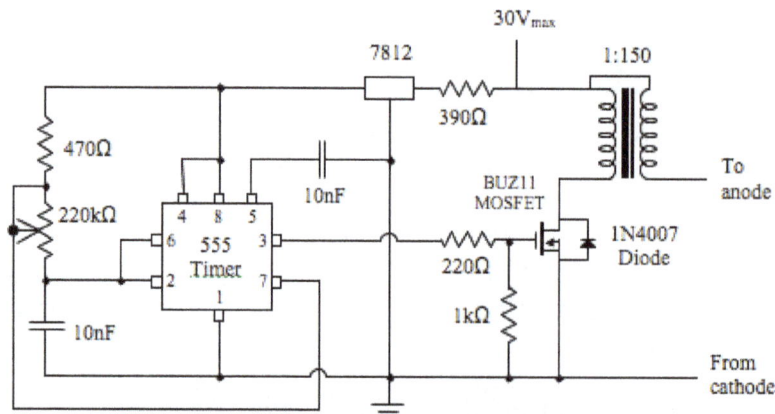

Driving circuits (E-type) of a power supply

The driving waveforms can be optimized to achieve a better actuation (induced flow speed). However, a sinusoidal waveform may be more preferable for the simplicity in power supply construction. The additional benefit is the relatively less electromagnetic interference. Pulse width modulation can be adopted to instantaneously adjust the strength of actuation.

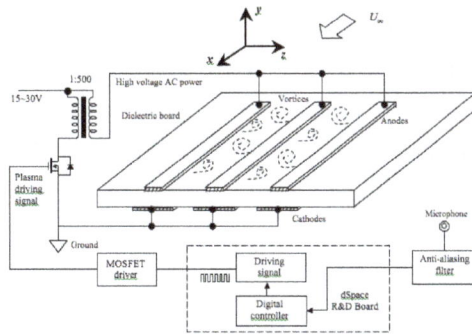

Pulse width modulation of plasma power input

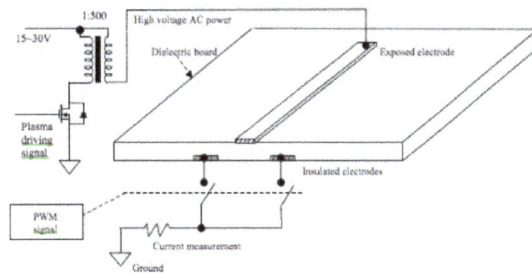

One configuration of DBD plasma actuator

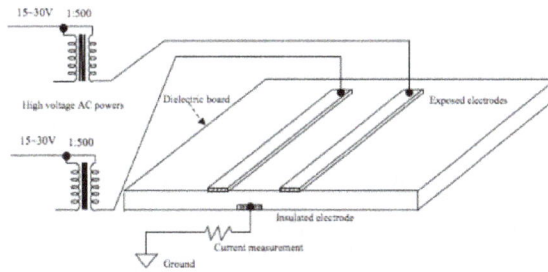

One configuration of DBD plasma actuator

Manipulation of the encapsulated electrode and distributing the encapsulated electrode throughout the dielectric layer has been shown to alter the performance of the dielectric barrier discharge (DBD) plasma actuator. Locating the initial encapsulated electrode closer to the dielectric surface results in induced velocities higher than the baseline case for a given voltage. In addition, Actuators with a shallow initial electrode are able to impart more momentum and mechanical power into the flow.

No matter how much funding has been invested and the number of various private claims of a high induced speed, the maximum, average speed induced by plasma actuators on an atmospheric pressure conviction, without any assistant of mechanical amplifier (chamber, cavity etc.), is still less than 10 m/s.

Influence of Temperature

When dealing with real life aircraft equipped with plasma actuators, it is important to consider the effect of temperature. The temperature variations encountered during a flight envelope may have adverse effects in actuator performance. It is found that for a constant peak-to-peak voltage the maximum velocity produced by the actuator depends directly on the dielectric surface temperature. The findings suggest that by changing the actuator temperature the performance can be maintained or even altered at different environmental conditions. Increasing dielectric surface temperature can increase the plasma actuator performance by increasing the momentum flux whilst consuming slightly higher energy.

Flow Control Applications

Some recent applications of plasma actuation include high-speed flow control using localized arc filament plasma actuators, and low-speed flow control using dielectric barrier discharges and sliding discharges. The present research of plasma actuators is mainly focused on three directions: (1) various designs of plasma actuators; (2) flow control applications; and (3) control-oriented modeling of flow applications under plasma actuation. In addition, new experimental and numerical methods are being developed to provide physical insights.

Vortex Generator

A plasma actuator induces a local flow speed perturbation, which will be developed downstream to a vortex sheet. As a result, plasma actuators can behave as vortex generators. The difference between this and traditional vortex generation is that there are no mechanical moving parts or any drilling holes on aerodynamic surfaces, demonstrating an important benefit of plasma actuators.

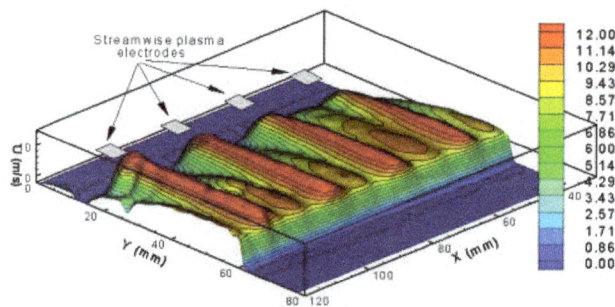

Plasma induced flow field

Active Noise Control

Active noise control normally denotes noise cancellation, that is, a noise-cancellation speaker emits a sound wave with the same amplitude but with inverted phase (also known as antiphase) to the original sound. However, active noise control with plasma adopts different strategies. The first one uses the discovery that sound pressure could be attenuated when it passes through a plasma sheet. The second one, and being more widely used, is to actively suppress the flow-field that is responsible to flow-induced noise (also known as aeroacoustics), using plasma actuators. It has been demonstrated that both tonal noise and broadband noise (difference can refer to tonal versus broadband) can be actively attenuated by a carefully designed plasma actuator.

Supersonic and Hypersonic Flow Control

Plasma has been introduced to hypersonic flow control. Firstly, plasma could be much easier generated for hypersonic vehicle at high altitude with quite low atmospheric pressure and high surface temperature. Secondly, the classical aerodynamic surface has little actuation for the case.

Interest in plasma actuators as active flow control devices is growing rapidly due to their lack of mechanical parts, light weight and high response frequency. The characteristics of a dielectric barrier discharge (DBD) plasma actuator when exposed to an unsteady flow generated by a shock tube is examined. A Study shows that not only is the shear layer outside of the shock tube affected by the plasma but the passage of the shock front and high-speed flow behind it also greatly influences the properties of the plasma

Flight Control

Plasma actuators could be mounted on the airfoil to control flight attitude and thereafter flight trajectory. The cumbersome design and maintenance efforts of mechanical and hydraulic transmission systems in a classical rudder can thus be saved. The price to pay is that one should design a suitable high voltage/power electric system satisfying EMC rule. Hence, in addition to flow control, plasma actuators hold potential in top-level flight control, in particular for UAV and extraterrestrial planet (with suitable atmospheric conditions) investigations.

On the other hand, the whole flight control strategy should be reconsidered taking account of characteristics of plasma actuators. One preliminary roll control system with DBD plasma actuators is shown in the figure.

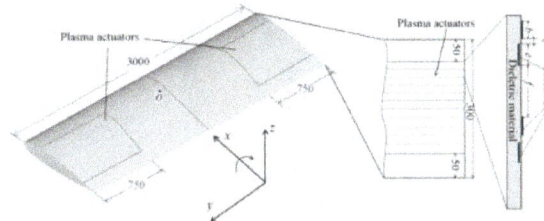

DBD Plasma actuators deployed on a NACA 0015 airfoil to do rudderless flight control

It can be seen that plasma actuators deployed on the both sides of an airfoil. The roll control can be controlled by activating plasma actuators according to the roll angle feedback. After studying various feedback control methodologies, the bang–bang control method was chosen to design the roll control system based on plasma actuators. The reason is that bang-bang control is time optimal and insensitive to plasma actuations, which quickly vary in difference atmospheric and electric conditions.

Modeling

Various numerical models have been proposed to simulate plasma actuations in flow control. They are listed below according to the computational cost, from the most expensive to the cheapest.

- Monte carlo method plus particle-in-cell;

- Electricity modeling coupled with Navier-Stokes equations;

- Lumped element model coupled with Navier-Stokes equations

- Surrogate model to simulate plasma actuation.

The most important potential of plasma actuators is its ability to bridge fluids and electricity. A modern closed-loop control system and the following information theoretical methods can be applied to the relatively classical aerodynamic sciences. A control-oriented model for plasma actuation in flow control has been proposed for a cavity flow control case.

Rotary Actuator

The simplest actuator is purely mechanical, where linear motion in one direction gives rise to rotation. The most common actuators though are electrically powered. Other actuators may be powered by pneumatic or hydraulic power, or may use energy stored internally through springs.

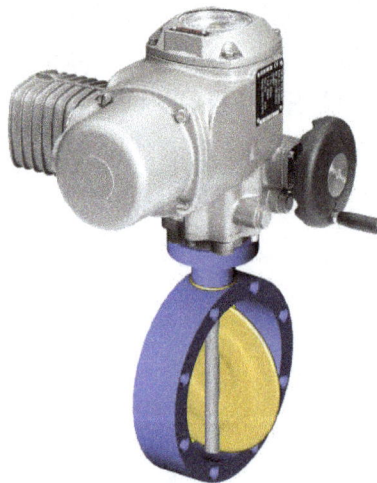

Electric rotary valve actuator controlling a butterfly valve

A rotary actuator is an actuator that produces a rotary motion or torque.

The motion produced by an actuator may be either continuous rotation, as for an electric motor, or movement to a fixed angular position as for servomotors and stepper motors. A further form, the torque motor, does not necessarily produce any rotation but merely generates a precise torque which then either causes rotation, or is balanced by some opposing torque.

Actuator Power Sources

Electric Actuators

Stepper Motors

Stepper motors are a form of electric motor that has the ability to move in discrete steps of a fixed size. This can be used either to produce continuous rotation at a controlled speed or to move by a controlled angular amount. If the stepper is combined with either a position encoder or at least a single datum sensor at the zero position, it is possible to move the motor to any angular position and so to act as a rotary actuator.

The variety of stepper motors

Servomotors

A servomotor is a packaged combination of several components: a motor (usually electric, although fluid power motors may also be used), a gear train to reduce the many rotations of the motor to a higher torque rotation, a position encoder that identifies the position of the output shaft and an inbuilt control system. The input control signal to the servo indicates the desired output position. Any difference between the position commanded and the position of the encoder gives rise to an error signal that causes the motor and geartrain to rotate until the encoder reflects a position matching that commanded.

Radio control servo

A simple low-cost servo of this type is widely used for radio-controlled models.

Other Types

A recent, and novel, form of ultra-lightweight actuator uses memory wire. As a current is applied, the wire is heated above its transition temperature and so changes shape, applying a torque to the output shaft. When power is removed, the wire cools and returns to its earlier shape.

Fluid Power Actuators

Both hydraulic and pneumatic power may be used to drive an actuator, usually the larger and more powerful types. As their internal construction is generally similar (in principle, if not in size) they are often considered together as *fluid power* actuators. Fluid power actuators are of two common forms: those where a linear piston and cylinder mechanism is geared to produce rotation (*illustrated*), and those where a rotating asymmetrical vane swings through a cylinder of two different radii. The differential pressure between the two sides of the vane gives rise to an unbalanced force and thus a torque on the output shaft. Vane actuators require a number of sliding seals and the joins between these seals have tended to cause more problems with leakage than for the piston and cylinder type.

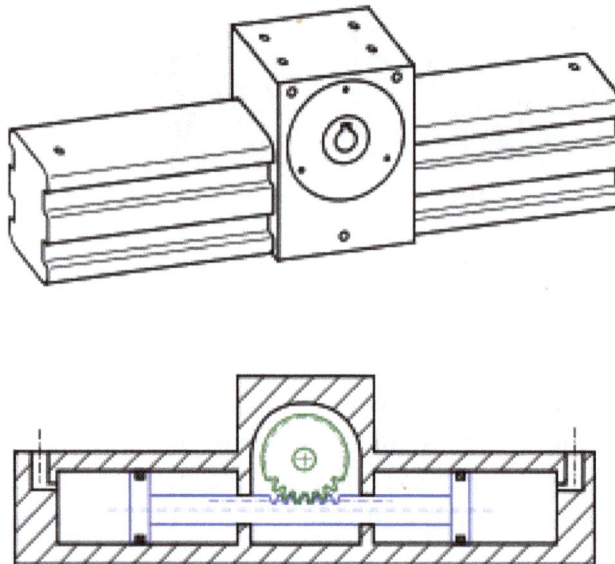

Hydraulic or pneumatic rotary actuator, using a rack and pinion

Vacuum Actuators

Where a supply of vacuum is available, but not pneumatic power, rotary actuators have even been made to work from vacuum power. The only common instance of these was for early automatic windscreen wipers on cars up until around 1960. These used the manifold vacuum of a petrol engine to work a quarter-turn oscillating vane actuator. Such windscreen wipers worked adequately when the engine was running under light load, but they were notorious that when working hard at top speed or climbing a hill, the manifold vacuum was reduced and the wipers slowed to a crawl.

Applications

Rotary actuators are used in a vast range of applications. These require actuators of all sizes, power and operating speed. These can range from zero power actuators that are only used as display devices, such as air core gauges. Others include valve actuators that operate pipeline and process valves in the petrochemical industry, through to actuators for large civil engineering projects such as sluice gates and dams.

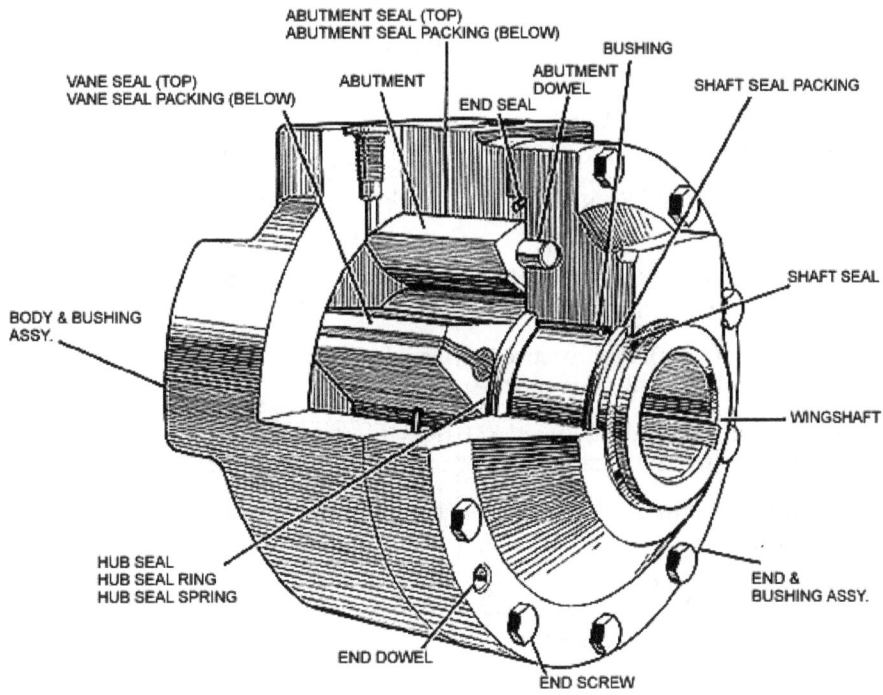

VANE SEAL (TOP)
VANE SEAL PACKING (BELOW)

ABUTMENT SEAL (TOP)
ABUTMENT SEAL PACKING (BELOW)

ABUTMENT

END SEAL

ABUTMENT
DOWEL

BUSHING

SHAFT SEAL PACKING

BODY & BUSHING
ASSY.

SHAFT SEAL

WINGSHAFT

HUB SEAL
HUB SEAL RING
HUB SEAL SPRING

END DOWEL

END SCREW

END &
BUSHING ASSY.

References

- Parker, Dana T. Building Victory: Aircraft Manufacturing in the Los Angeles Area in World War II, p. 87, Cypress, CA, 2013. ISBN 978-0-9897906-0-4.

- Carlisle, Rodney (2004). Scientific American Inventions and Discoveries, p. 266. John Wiley & Sons, Inc., New Jersey. ISBN 0-471-24410-4.

- De la Vergne, Jack (2003). Hard Rock Miner's Handbook. Tempe/North Bay: McIntosh Engineering. pp. 114–124. ISBN 0-9687006-1-6.

- Cowie, J.M.G.; Valerai Arrighi (2008). "13". Polymers: Chemistry and Physics of Modern Material (Third ed.). Florida: CRC Press. pp. 363–373. ISBN 978-0-8493-9813-1.

- Kim, K.J.; Tadokoro, S. (2007). Electroactive Polymers for Robotic Applications, Artificial Muscles and Sensors. London: Springer. ISBN 978-1-84628-371-0.

- Glass, J. Edward; Schulz, Donald N.; Zukosi, C.F (May 13, 1991). "1". Polymers as Rheology Modifiers. ACS Symposium Series. 462. Americal Chemical Society. pp. 2–17. ISBN 9780841220096.

- "Advantages of Hydraulic Presses". MetalFormingFacts.com. The Lubrizol Corporation. 4 February 2013. Retrieved 23 August 2013.

Actuators: Technologies and Devices

The chapter serves as a source to understand the technologies and devices of actuators. Some of the technologies and devices involved in actuators are hydraulic press, jackscrew, roller screw, and pneumatic cylinder. Tools and techniques are an important component of any field of study. The following chapter elucidates the various tools and techniques that are related to actuator technology.

Pneumatic Motor

A pneumatic motor (Air motor) or compressed air engine is a type of motor which does mechanical work by expanding compressed air. Pneumatic motors generally convert the compressed air energy to mechanical work through either linear or rotary motion. Linear motion can come from either a diaphragm or piston actuator, while rotary motion is supplied by either a vane type air motor, piston air motor, air turbine or gear type motor.

The Victor Tatin airplane of 1879 used a compressed-air engine for propulsion. Original craft, at Musée de l'Air et de l'Espace.

Pneumatic motors have existed in many forms over the past two centuries, ranging in size from hand-held motors to engines of up to several hundred horsepower. Some types rely on pistons and cylinders; others on slotted rotors with vanes (vane motors) and other uses turbines. Many compressed air engines improve their performance by heating the incoming air or the engine itself. Pneumatic motors have found widespread success in the hand-held tool industry, but are also used stationary in a wide range of industrial applications. Continual attempts are being made to expand their use to the transportation industry. However, pneumatic motors must overcome inefficiencies before being seen as a viable option in the transportation industry.

The first mechanically powered submarine, the 1863 *French submarine Plongeur*, used a compressed-air engine.
Musée de la Marine (Rochefort).

Classification

Linear

In order to achieve linear motion from compressed air, a system of pistons is most commonly used. The compressed air is fed into an air-tight chamber that houses the shaft of the piston. Also inside this chamber a spring is coiled around the shaft of the piston in order to hold the chamber completely open when air is not being pumped into the chamber. As air is fed into the chamber the force on the piston shaft begins to overcome the force being exerted on the spring. As more air is fed into the chamber, the pressure increases and the piston begins to move down the chamber. When it reaches its maximum length the air pressure is released from the chamber and the spring completes the cycle by closing off the chamber to return to its original position.

Piston motors are the most commonly used in hydraulic systems. Essentially, piston motors are the same as hydraulic motors except they are used to convert hydraulic energy into mechanical energy.

Piston motors are often used in series of two, three, four, five, or six cylinders that are enclosed in a housing. This allows for more power to be delivered by the pistons because several motors are in sync with each other at certain times of their cycle. The above-mentioned technique for generating the electricity through compressed air can be implemented in a weaving air jet unit.

Rotary Vane Motors

A type of pneumatic motor, known as a rotary vane motor, uses air to produce rotational motion to a shaft. The rotating element is a slotted rotor which is mounted on a drive shaft. Each slot of the rotor is fitted with a freely sliding rectangular vane. The vanes are extended to the housing walls using springs, cam action, or air pressure, depending on the motor design. Air is pumped through the motor input which pushes on the vanes creating the rotational motion of the central shaft. Rotation speeds can vary between 100 and 25,000 rpm depending on several factors which include the amount of air pressure at the motor inlet and the diameter of the housing.

One application for vane-type air motors is to start large industrial diesel or natural gas engines. Stored energy in the form of compressed air, nitrogen or natural gas enters the sealed motor chamber and exerts pressure against the vanes of a rotor. This causes the rotor to turn at high speed.

Because the engine flywheel requires a great deal of torque to start the engine, reduction gears are used. Reduction gears create high torque levels with the lower amounts of energy input. These reduction gears allow for sufficient torque to be generated by the engine flywheel while it is engaged by the pinion gear of the air motor or air starter.

Application

A widespread application of pneumatic motors is in hand-held tools, impact wrenches, pulse tools, screwdrivers, nut runners, drills, grinders, sanders and so on. Pneumatic motors are also used stationary in a wide range of industrial applications. Though overall energy efficiency of pneumatics tools is low and they require access to a compressed-air source, there are several advantages over electric tools. They offer greater power density (a smaller pneumatic motor can provide the same amount of power as a larger electric motor), do not require an auxiliary speed controller (adding to its compactness), generate less heat, and can be used in more volatile atmospheres as they do not require electric power and do not create sparks. They can be loaded to stop with full torque without damages.

Historically, many individuals have tried to apply pneumatic motors to the transportation industry. Guy Negre, CEO and founder of Zero Pollution Motors, has pioneered this field since the late 1980s. Recently Engineair has also developed a rotary motor for use in automobiles. Engineair places the motor immediately beside the wheel of the vehicle and uses no intermediate parts to transmit motion which means almost all of the motor's energy is used to rotate the wheel.

History in Transportation

The pneumatic motor was first applied to the field of transportation in the mid-19th century. Though little is known about the first recorded compressed-air vehicle, it is said that the Frenchmen Andraud and Tessie of Motay ran a car powered by a pneumatic motor on a test track in Chaillot, France, on July 9, 1840. Although the car test was reported to have been successful, the pair didn't explore further expansion of the design.

The first successful application of the pneumatic motor in transportation was the Mekarski system air engine used in locomotives. Mekarski's innovative engine overcame cooling that accompanies air expansion by heating air in a small boiler prior to use. The Tramway de Nantes, located in Nantes, France, was noted for being the first to use Mekarski engines to power their fleet of locomotives. The tramway began operation on December 13, 1879, and continues to operate today, although the pneumatic trams were replaced in 1917 by more efficient and modern electrical trams.

American Charles Hodges also found success with pneumatic motors in the locomotive industry. In 1911 he designed a pneumatic locomotive and sold the patent to the H.K. Porter Company in Pittsburgh for use in coal mines. Because pneumatic motors do not use combustion they were a much safer option in the coal industry.

Many companies claim to be developing Compressed air cars, but none are actually available for purchase or even independent testing.

Tools

Impact wrenches, pulse tools, nutrunners, screwdrivers, drills, grinders, die grinders, sanders, dental drills and other pneumatic tools use a variety of air motors. These include vane type motors, turbines and piston motors.

Torpedoes

Most successful early forms of self-propelled torpedoes used high pressure compressed air, although this was superseded by internal or external combustion engines, steam engines, or electric motors.

Railways

Compressed air engines were used in trams and shunters, and eventually found a successful niche in mining locomotives, although in the end they were replaced by electric trains, underground. Over the years designs increased in complexity, resulting in a triple expansion engine with air-to-air reheaters between each stage.

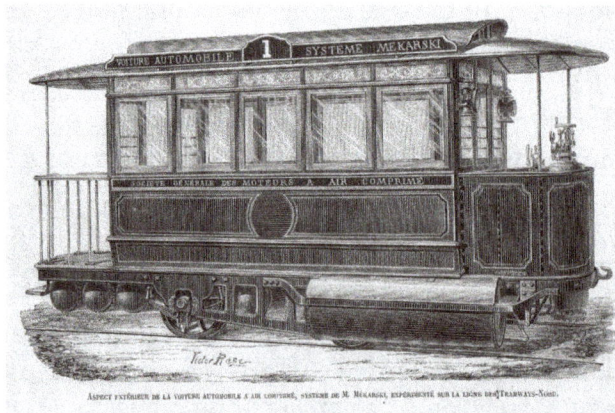

Mekarski compressed air tram, 1875

Pneumatic Locomotive with attached pressure container used during the construction of the Gotthard Rail Tunnel 1872-1880.

Flight

Transport category airplanes, such as commercial airliners, use compressed air starters to start the main engines. The air is supplied by the load compressor of the aircraft's auxiliary power unit, or by ground equipment.

Water rockets use compressed air to power their water jet and generate thrust, they are used as toys.

Air Hogs, a toy brand, also uses compressed air to power piston engines in toy airplanes (and some other toy vehicles).

Automotive

There is currently some interest in developing air cars. Several engines have been proposed for these, although none have demonstrated the performance and long life needed for personal transport.

Energine

The Energine Corporation was a South Korean company that claimed to deliver fully assembled cars running on a hybrid compressed air and electric engine. The compressed-air engine is used to activate an alternator, which extends the autonomous operating capacity of the car. The CEO was arrested for fraudulently promoting air motors with false claims.

EngineAir

EngineAir, an Australian company, is making a rotary engine powered by compressed air, called The Di Pietro motor. The Di Pietro motor concept is based on a rotary piston. Different from existing rotary engines, the Di Pietro motor uses a simple cylindrical rotary piston (shaft driver) which rolls, with little friction, inside the cylindrical stator.

It can be used in boat, cars, burden carriers and other vehicles. Only 1 psi ($\approx$ 6,8 kPa) of pressure is needed to overcome the friction. The engine was also featured on the ABC's New Inventors programme in Australia on 24 March 2004.

K'Airmobiles

K'Airmobiles vehicles were intended to be commercialized from a project developed in France in 2006-2007 by a small group of researchers. However, the project has not been able to gather the necessary funds.

People should note that, meantime, the team has recognized the physical impossibility to use onboard stored compressed air due to its poor energy capacity and the thermal losses resulting from the expansion of the gas.

These days, using the patent pending 'K'Air Generator', converted to work as a compressed-gas motor, the project should be launched in 2010, thanks to a North American group of investors, but for the purpose of developing first a green energy power system.

MDI

In the original Nègre air engine, one piston compresses air from the atmosphere to mix with the stored compressed air (which will cool drastically as it expands). This mixture drives the second piston, providing the actual engine power. MDI's engine works with constant torque, and the only way to change the torque to the wheels is to use a pulley transmission of constant variation, losing some efficiency. When vehicle is stopped, MDI's engine had to be on and working, losing energy. In 2001-2004 MDI switched to a design similar to that described in Regusci's patents, which date back to 1990.

It has been reported in 2008 that Indian car manufacturer Tata was looking at an MDI compressed air engine as an option on its low priced Nano automobiles. Tata announced in 2009 that the compressed air car was proving difficult to develop due to its low range and problems with low engine temperatures.

Quasiturbine

The Pneumatic Quasiturbine engine is a compressed air pistonless rotary engine using a rhomboidal-shaped rotor whose sides are hinged at the vertices.

The Quasiturbine has demonstrated as a pneumatic engine using stored compressed air

It can also take advantage of the energy amplification possible from using available external heat, such as solar energy.

The Quasiturbine rotates from pressure as low as 0.1 atm (1.47psi).

Since the Quasiturbine is a pure expansion engine, while the Wankel and most other rotary engines are not, it is well-suited as a compressed fluid engine, air engine or air motor.

Regusci

Armando Regusci's version of the air engine couples the transmission system directly to the wheel, and has variable torque from zero to the maximum, enhancing efficiency. Regusci's patents date from 1990.

Team Psycho-active

Psycho-Active is developing a multi-fuel/air-hybrid chassis which is intended to serve as the foundation for a line of automobiles. Claimed performance is 50 hp/litre. The compressed air motor they use is called the DBRE or Ducted Blade Rotary Engine.

Defunct Air Engine Designs

Conger Motor

Milton M. Conger in 1881 patented and supposedly built a motor that ran off compressed air or steam that using a flexible tubing which will form a wedge-shaped or inclined wall or abutment in the rear of the tangential bearing of the wheel, and propel it with greater or less speed according to the pressure of the propelling medium.

Pneumatic Cylinder

Pneumatic cylinder(s) (sometimes known as air cylinders) are mechanical devices which use the power of compressed gas to produce a force in a reciprocating linear motion.

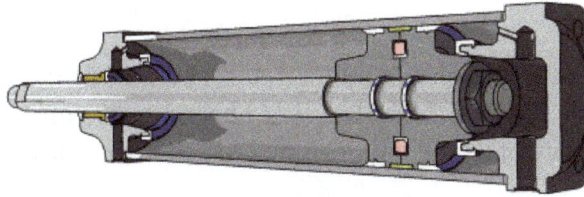

3D animated pneumatic cylinder (CAD)

Like hydraulic cylinders, something forces a piston to move in the desired direction. The piston is a disc or cylinder, and the piston rod transfers the force it develops to the object to be moved. Engineers sometimes prefer to use pneumatics because they are quieter, cleaner, and do not require large amounts of space for fluid storage.

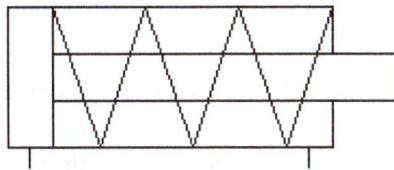

Schematic symbol for pneumatic cylinder with spring return

Because the operating fluid is a gas, leakage from a pneumatic cylinder will not drip out and contaminate the surroundings, making pneumatics more desirable where cleanliness is a requirement. For example, in the mechanical puppets of the Disney Tiki Room, pneumatics are used to prevent fluid from dripping onto people below the puppets.

Operation

General

Once actuated, compressed air enters into the tube at one end of the piston and, hence, imparts force on the piston. Consequently, the piston becomes displaced.

Compressibility of Gasses

One major issue engineers come across working with pneumatic cylinders has to do with the compressibility of a gas. Many studies have been completed on how the precision of a pneumatic cylinder can be affected as the load acting on the cylinder tries to further compress the gas used. Under a vertical load, a case where the cylinder takes on the full load, the precision of the cylinder is affected the most. A study at the National Cheng Kung University in Taiwan, concluded that the accuracy is about ± 30 nm, which is still within a satisfactory range but shows that the compressibility of air has an effect on the system.

Fail Safe Mechanisms

Pneumatic systems are often found in settings where even rare and brief system failure is unacceptable. In such situations locks can sometimes serve as a safety mechanism in case of loss of air supply (or its pressure falling) and, thus remedy or abate any damage arising in such a situation. Leakage of air from the input or output reduces the pressure and so the desired output.

Types

Although pneumatic cylinders will vary in appearance, size and function, they generally fall into one of the specific categories shown below. However, there are also numerous other types of pneumatic cylinder available, many of which are designed to fulfill specific and specialized functions.

Single-acting Cylinders

Single-acting cylinders (SAC) use the pressure imparted by compressed air to create a driving force in one direction (usually out), and a spring to return to the "home" position. More often than not, this type of cylinder has limited extension due to the space the compressed spring takes up. Another downside to SACs is that part of the force produced by the cylinder is lost as it tries to push against the spring.

Double-acting Cylinders

Double-acting cylinders (DAC) use the force of air to move in both extend and retract strokes. They have two ports to allow air in, one for outstroke and one for instroke. Stroke length for this design is not limited, however, the piston rod is more vulnerable to buckling and bending. Additional calculations should be performed as well.

Multi-stage, Telescoping Cylinder

Telescoping cylinders, also known as telescopic cylinders can be either single or double-acting. The telescoping cylinder incorporates a piston rod nested within a series of hollow stages of increasing diameter. Upon actuation, the piston rod and each succeeding stage "telescopes" out as a segmented piston. The main benefit of this design is the allowance for a notably longer stroke than would be achieved with a single-stage cylinder of the same collapsed (retracted) length. One cited drawback to telescoping cylinders is the increased potential for piston flexion due to the segmented piston design. Consequently, telescoping cylinders are primarily utilized in applications where the piston bears minimal side loading.

pneumatic telescoping cylinder, 8-stages, single-acting, retracted and extended

Other Types

Although SACs and DACs are the most common types of pneumatic cylinder, the following types are not particularly rare:

- Through rod air cylinders: piston rod extends through both sides of the cylinder, allowing for equal forces and speeds on either side.

- Cushion end air cylinders: cylinders with regulated air exhaust to avoid impacts between the piston rod and the cylinder end cover.

- Rotary air cylinders: actuators that use air to impart a rotary motion.

- Rodless air cylinders: These have no piston rod. They are actuators that use a mechanical or magnetic coupling to impart force, typically to a table or other body that moves along the length of the cylinder body, but does not extend beyond it.

- Tandem air cylinder: two cylinders are assembled in series in order to double the force output.

- Impact air cylinder: high velocity cylinders with specially designed end covers that withstand the impact of extending or retracting piston rods.

Rodless Cylinders

Some rodless types have a slot in the wall of the cylinder that is closed off for much of its length by two flexible metal sealing bands. The inner one prevents air from escaping, while the outer one protects the slot and inner band. The piston is actually a pair of them, part of a comparatively long assembly. They seal to the bore and inner band at both ends of the assembly. Between the individual pistons, however, are camming surfaces that "peel off" the bands as the whole sliding assembly moves toward the sealed volume, and "replace" them as the assembly moves away from the other end. Between the camming surfaces is part of the moving assembly that protrudes through the slot to move the load. Of course, this means that the region where the sealing bands are not in contact is at atmospheric pressure.

Another type has cables (or a single cable) extending from both (or one) end[s] of the cylinder. The cables are jacketed in plastic (nylon, in those referred to), which provides a smooth surface that permits sealing the cables where they pass through the ends of the cylinder. Of course, a single cable has to be kept in tension.

Still others have magnets inside the cylinder, part of the piston assembly, that pull along magnets outside the cylinder wall. The latter are carried by the actuator that moves the load. The cylinder wall is thin, to ensure that the inner and outer magnets are near each other. Multiple modern high-flux magnet groups transmit force without disengaging or excessive resilience.

Design

Construction

Depending on the job specification, there are multiple forms of body constructions available:

- Tie rod cylinders: The most common cylinder constructions that can be used in many types of loads. Has been proven to be the safest form.

- Flanged-type cylinders: Fixed flanges are added to the ends of cylinder, however, this form of construction is more common in hydraulic cylinder construction.

- One-piece welded cylinders: Ends are welded or crimped to the tube, this form is inexpensive but makes the cylinder non-serviceable.

- Threaded end cylinders: Ends are screwed onto the tube body. The reduction of material can weaken the tube and may introduce thread concentricity problems to the system.

Material

Upon job specification, the material may be chosen. Material range from nickel-plated brass to aluminum, and even steel and stainless steel. Depending on the level of loads, humidity, temperature, and stroke lengths specified, the appropriate material may be selected.

Mounts

Depending on the location of the application and machinability, there exist different kinds of mounts for attaching pneumatic cylinders:

Type of Mount Ends	
Rod End	**Cylinder End**
Plain	Plain
Threaded	Foot
Clevis	Bracket-single or double
Torque or eye	Trunnion
Flanged	Flanged
	Clevis etc.

Sizes

Air cylinders are available in a variety of sizes and can typically range from a small 2.5 mm ($\frac{1}{10}$ in) air cylinder, which might be used for picking up a small transistor or other electronic component, to 400 mm (16 in) diameter air cylinders which would impart enough force to lift a car. Some pneumatic cylinders reach 1,000 mm (39 in) in diameter, and are used in place of hydraulic cylinders for special circumstances where leaking hydraulic oil could impose an extreme hazard.

Pressure, Radius, Area and Force Relationships

Rod Stresses

Due to the forces acting on the cylinder, the piston rod is the most stressed component and has to be designed to withstand high amounts of bending, tensile and compressive forces. Depending on how long the piston rod is, stresses can be calculated differently. If the rods length is less than 10 times the diameter, then it may be treated as a rigid body which has compressive or tensile forces acting on it. In which case the relationship is:

$$F = A\sigma$$

Where:

F is the compressive or tensile force

A is the cross-sectional area of the piston rod

σ is the stress

However, if the length of the rod exceeds the 10 times the value of the diameter, then the rod needs to be treated as a column and buckling needs to be calculated as well.

Instroke and Outstroke

Although the diameter of the piston and the force exerted by a cylinder are related, they are not directly proportional to one another. Additionally, the typical mathematical relationship between the two assumes that the air supply does not become saturated. Due to the effective cross sectional area reduced by the area of the piston rod, the instroke force is less than the outstroke force when both are powered pneumatically and by same supply of compressed gas.

The relationship between the force, radius, and pressure can derived from simple distributed load equation:

$$F_r = PA_e$$

Where:

F_r is the resultant force

P is the pressure or distributed load on the surface

A_e is the effective cross sectional area the load is acting on

Outstroke

Using the distributed load equation provided the A_e can be replaced with area of the piston surface where the pressure is acting on.

$$F_r = P(\pi r^2)$$

Where:

F_r represents the resultant force

r represents the radius of the piston

π is pi, approximately equal to 3.14159.

Instroke

On instroke, the same relationship between force exerted, pressure and *effective cross sectional area* applies as discussed above for outstroke. However, since the cross sectional area is less than the piston area the relationship between force, pressure and *radius* is different. The calculation isn't more complicated though, since the effective cross sectional area is merely that of the piston surface minus the cross sectional area of the piston rod.

For instroke, therefore, the relationship between force exerted, pressure, radius of the piston, and radius of the piston rod, is as follows:

$$F_r = P(\pi r_1^2 - \pi r_2^2) = P\pi(r_1^2 - r_2^2)$$

Where:

F_r represents the resultant force

r_1 represents the radius of the piston

r_2 represents the radius of the piston rod

π is pi, approximately equal to 3.14159.

Hydraulic Press

A hydraulic press is a device using a hydraulic cylinder to generate a compres-sive force. It uses the hydraulic equivalent of a mechanical lever, and was also known as a Bramah press after the inventor, Joseph Bramah, of England. He invented and was issued a patent on this press in 1795. As Bramah (who is also known for his development of the flush toilet) installed toilets, he studied the existing literature on the motion of fluids and put this knowledge into the development of the press.

Hydraulic press - 400T

$$F2 = F1\ (A2/A1)$$

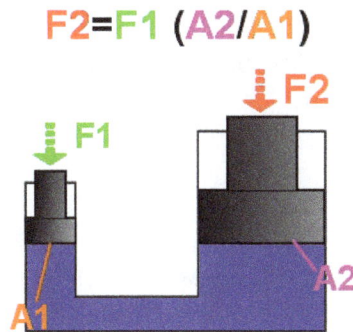

Hydraulic force increase

Principle

The hydraulic press depends on Pascal's principle: the pressure throughout a closed system is constant. One part of the system is a piston acting as a pump, with a modest mechanical force acting on a small cross-sectional area; the other part is a piston with a larger area which generates a correspondingly large mechanical force. Only small-diameter tubing (which more easily resists pressure) is needed if the pump is separated from the press cylinder.

Pascal's law: Pressure on a confined fluid is transmitted undiminished and acts with equal force on equal areas and at 90 degrees to the container wall.

Hydraulic Press in a machine shop. This press is commonly used for hydroforming.

A fluid, such as oil, is displaced when either piston is pushed inward. Since the fluid is incompressible, the volume that the small piston displaces is equal to the volume displaced by the large piston. This causes a difference in the length of displacement, which is proportional to the ratio of areas of the heads of the pistons, given that volume = area × length. Therefore, the small piston must be moved a large distance to get the large piston to move significantly. The distance the large piston will move is the distance that the small piston is moved divided by the ratio of the areas of the heads of the pistons. This is how energy, in the form of work in this case, is conserved and the Law of Conservation of Energy is satisfied. Work is force applied over a distance, and since the force is increased on the larger piston, the distance the force is applied over must be decreased.

Bramah's basic idea is also exploited in hydroforming.

Application

Hydraulic presses are commonly used for forging, clinching, moulding, blanking, punching, deep drawing, and metal forming operations.

In Popular Culture

The room featured in *Fermat's Room* has a design similar to that of a hydraulic press. Boris Artzybasheff also created a drawing of a hydraulic press, in which the press was created out of the shape of a robot. In 2015, the Hydraulic Press Channel, a YouTube channel solely dedicated to crushing objects with a hydraulic press, was created and quickly rose to fame. The channel is owned and operated by Lauri Vuohensilta from Tampere, Finland.

Jackscrew

A jackscrew is a type of jack that is operated by turning a leadscrew. In the form of a screw jack it is commonly used to lift moderately heavy weights, such as vehicles. More commonly it is used as an adjustable support for heavy loads, such as the foundations of houses, or large vehicles. These can support a heavy load, but not lift it.

A 2.5-ton screw jack. The jack is operated by inserting the bar *(visible lower left)* in the holes at the top and turning.

A jackscrew operates this automotive scissor jack.

Antique locomotive screw jack

Antique wooden jackscrew for repair of cart and wagon wheels (Ethnographic Museum of Elhovo, Bulgaria)

Advantages

An advantage of jackscrews over some other types of jack is that they are *self-locking*, which means when the rotational force on the screw is removed, it will remain motionless where it was left and will not rotate backwards, regardless of how much load it is supporting. This makes them inherently safer than hydraulic jacks, for example, which will move backwards under load if the force on the hydraulic actuator is accidentally released.

Mechanical Advantage

The mechanical advantage of a screw jack, the ratio of the force the jack exerts on the load to the input force on the lever, ignoring friction is

$$\frac{F_{load}}{F_{in}} = \frac{2\pi r}{l}$$

where

F_{load} is the force the jack exerts on the load

F_{in} is the rotational force exerted on the handle of the jack

r is the length of the jack handle, from the screw axis to where the force is applied

l is the lead of the screw.

This derives from two factors, the simple lever advantage of a long operating handle and also the advantage of the inclined plane of the leadscrew. However, most screw jacks have large amounts of friction which increase the input force necessary, so the actual mechanical advantage is often only 30% to 50% of this figure.

Limitations

Screw jacks are limited in their lifting capacity. Increasing load increases friction within the screw threads. A fine pitch thread, which would increase the advantage of the screw, also reduces the size and strength of the threads. Longer operating levers soon reach a point where the lever will simply bend at their inner end.

Screw jacks have now largely been replaced by hydraulic jacks. This was encouraged in 1858 when jacks by the Tangye company to Bramah's hydraulic press concept were applied to the successful launching of Brunel's SS *Great Britain*, after two failed attempts by other means. The maximum mechanical advantage possible for a hydraulic jack is not limited by the limitations on screw jacks and can be far greater. After WWII, improvements to the grinding of hydraulic rams and the use of O ring seals reduced the price of low-cost hydraulic jacks and they became widespread for use with domestic cars. Screw jacks still remain for minimal cost applications, such as the little-used tyre-changing jacks supplied with cars.

Applications

A jackscrew's threads must support heavy loads. In the most heavy-duty applications, such as screw jacks, a square thread or buttress thread is used, because it has the lowest friction. In other application such as actuators, an Acme thread is used, although it has higher friction.

The large area of sliding contact between the screw threads means jackscrews have high friction and low efficiency as power transmission linkages, around 30%–50%. So they are not often used for continuous transmission of high power, but more often in intermittent positioning applications.

The ball screw is a more advanced type of leadscrew that uses a recirculating-ball nut to minimize friction and prolong the life of the screw threads. The thread profile of such screws is approximately semicircular (commonly a "gothic arch" profile) to properly mate with the bearing balls. The disadvantage to this type of screw is that it is not self-locking. Ball screws are prevalent in powered leadscrew actuators.

Jackscrews form vital components in equipment. For instance, the failure of a jackscrew on a McDonnell Douglas MD80 airliner due to a lack of grease resulted in the crash of Alaska Airlines Flight 261 off the coast of California in 2000.

The jackscrew figured prominently in the classic novel *Robinson Crusoe*. It was also featured in a recent History Channel program as *the* saving tool of the Pilgrims' voyage – the main crossbeam, a key structural component of their small ship, cracked during a severe storm. A farmer's jackscrew secured the damage until landfall.

In Electronic Connectors

The term *jackscrew* is also used for the captive screws that draw the two parts of D-subminiature electrical connectors together and hold them mated. When unscrewed, they allow the connector halves to be taken apart. These small jackscrews may have ordinary screw heads or extended heads (also making them thumbscrews) that allow the user's fingers to turn the jackscrew. Furthermore, the head sometimes has an internal female thread, with the male externally threaded screw shaft extending from that. The threaded-head type can be used to panel-mount one connector and provide a means to attach the mating connector to the first connector.

Hoist (Device)

A hoist is a device used for lifting or lowering a load by means of a drum or lift-wheel around which rope or chain wraps. It may be manually operated, electrically or pneumatically driven and may use chain, fiber or wire rope as its lifting medium. The load is attached to the hoist by means of a lifting hook.

Types of Hoist

Builder's hoist, with small gasoline engine

The basic hoist has two important characteristics to define it: Lifting medium and power type. The lifting medium is either wire rope, wrapped around a drum, or load-chain, raised by a pulley with a special profile to engage the chain. The power can be provided by different means. Common means are hydraulics, electrical and air driven motors. Both the wire rope hoist and chain hoist have been in common use since the 1800s, however mass production of an electric hoist did not start until the early 1900s and was first adapted by Germany. A hoist can be built as one integral-package unit, designed for cost-effective purchasing and moderate use, or it can be built as a built-up custom unit, designed for durability and performance. The built-up hoist will be much more expensive, but will also be easier to repair and more durable. Package units were once regarded as being designed for light to moderate usage, but since the 60s this has changed. Built-up units are designed for heavy to severe service, but over the years that market has decreased in size since the advent of the more durable packaged hoist. A machine shop or fabricating shop will use an integral-package hoist, while a Steel Mill or NASA would use a built-up unit to meet durability, performance, and repairability requirements. NASA has also seen a change in the use of package hoists. The NASA Astronaut training pool, for example, utilizes cranes with packaged hoists.

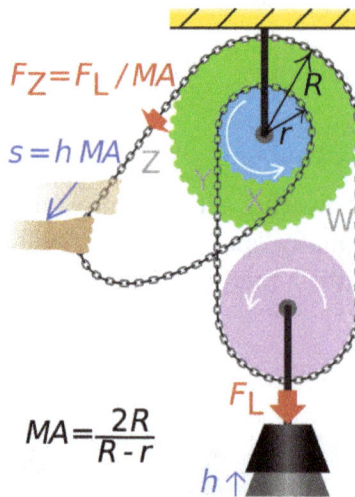

$$F_Z = F_L / MA$$

$$s = h\,MA$$

$$MA = \frac{2R}{R-r}$$

$$F_L$$

$$h \uparrow$$

A differential pulley chain hoist

A hoist on the Trump International Hotel & Tower-Chicago

Wire Rope Hoist or Chain Hoist

The more commonly used hoist in today's worldwide market is an electrically powered hoist. These are either the chain type or the wire rope type.

Nowadays many hoists are package hoists, built as one unit in a single housing, generally designed for ten-year life, but the life calculation is based on an industry standard when calculating actual life. See the Hoists Manufacturers Institute site for true life calculation which is based on load and hours used. In today's modern world for the North American market there are a few governing bodies for the industry. The Overhead Alliance is a group that represents Crane Manufacturers Association of America (CMAA), Shanghai WANBO Hoisting Machinery(VANBON), Hoist Manufacturers Institute (HMI), and Monorail Manufacturers Association (MMA). These product counsels of the Material Handling Industry of America have joined forces to create promotional materials to raise the awareness of the benefits to overhead lifting. The members of this group are marketing representatives of the member companies.

Hoist by Leonardo da Vinci (*Codex Atlanticus*, foglio 30v., 1480 ca.). Reconstruction exhibited at the Museo nazionale della scienza e della tecnologia Leonardo da Vinci, Milan.

Common small portable hoists are of two main types, the *chain hoist* or *chain block* and the wire rope or cable type. Chain hoists may have a lever to actuate the hoist or have a loop of operating chain that one pulls through the block (known traditionally as a chain fall) which then activates the block to take up the main lifting chain.

For a given *rated load* wire rope is lighter in weight per unit length but overall length is limited by the drum diameter that the cable must be wound onto. The lift chain of a chain hoist is far larger than the liftwheel over which chain may function. Therefore, a high-performance chain hoist may be of significantly smaller physical size than a wire rope hoist rated at the same working load.

Both systems fail over time through fatigue fractures if operated repeatedly at loads more than a small percentage of their tensile breaking strength. Hoists are often designed with internal clutches to limit operating loads below this threshold. Within such limits wire rope rusts from the inside outward while chain links are markedly reduced in cross section through wear on the inner surfaces. Regular lubrication of both tensile systems is recommended to reduce frequency of replacement. High speed lifting, greater than about 60 feet per minute (18.3 m/min), requires wire rope wound on a drum, because chain over a pocket wheel generates fatigue-inducing resonance for long lifts.

The unloaded wire rope of small hand-powered hoists often exhibits a snarled "set", making the use of a chain hoist in this application less frustrating, but heavier. In addition, if the wire in a wire hoist fails, it can whip and cause injury, while a chain will simply break.

"Chain hoist" also describes a hoist using a differential pulley system, in which a compound pulley with two different radii and teeth engage an endless chain, allowing the exerted force to be multiplied according to the ratio of the radii.

Construction Hoists

Also known as a Man-Lift, Buckhoist, temporary elevator, builder hoist, passenger hoist or construction elevator, this type of hoist is commonly used on large scale construction projects, such as high-rise buildings or major hospitals. There are many other uses for the construction elevator. Many other industries use the buckhoist for full-time operations, the purpose being to carry personnel, materials, and equipment quickly between the ground and higher floors, or between floors in the middle of a structure. There are three types: Utility to move material, personnel to move personnel, and dual-rated, which can do both.

The construction hoist is made up of either one or two cars (cages) which travel vertically along stacked mast tower sections. The mast sections are attached to the structure or building every 25 feet (7.62 m) for added stability. For precisely controlled travel along the mast sections, modern construction hoists use a motorized rack-and-pinion system that climbs the mast sections at various speeds.

While hoists have been predominantly produced in Europe and the United States, China is emerging as a manufacturer of hoists to be used in Asia.

In the United States and abroad, General Contractors and various other industrial markets rent or lease hoists for a specific projects. Rental or leasing companies provide erection, dismantling, and repair services to their hoists to provide General Contractors with turnkey services. Also, the rental and leasing companies can provide parts and service for the elevators that are under contract.

Mine Hoists

In underground mining a hoist or winder is used to raise and lower conveyances within the mine shaft. Human, animal and water power were used to power the mine hoists documented in Agricola's De Re Metallica, published in 1556. Stationary steam engines were commonly used to power mine hoists through the 19th century and into the 20th, as at the Quincy Mine, where a 4-cylinder cross-compound corliss engine was used. Modern hoists are powered using electric motors, historically with direct current drives utilizing solid-state converters (thyristors); however, modern large hoists use alternating current drives that are variable-frequency controlled. There are three principal types of hoists used in mining applications, Drum Hoists, Friction (or Kope) hoists and Blair multi-rope hoists.Hoist can be defined as anything that used to lift any heavy materials.

Electroactive Polymers

Electroactive polymers, or EAPs, are polymers that exhibit a change in size or shape when stimulated by an electric field. The most common applications of this type of material are in actuators and sensors. A typical characteristic property of an EAP is that they will undergo a large amount of deformation while sustaining large forces.

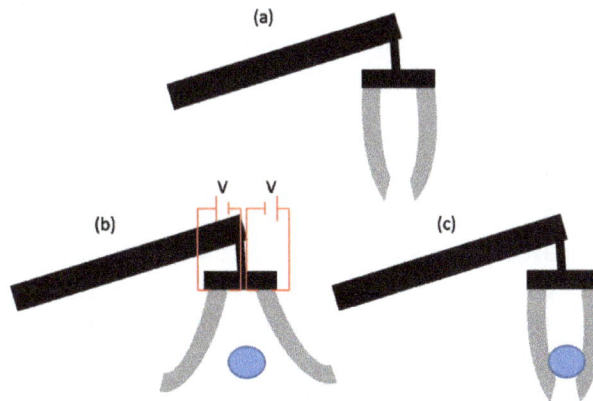

(a) Cartoon drawing of an EAP gripping device.
(b) A voltage is applied and the EAP fingers deform in order to surround the ball.
(c) When the voltage is removed the EAP fingers return to their original shape and grip the ball.

The majority of historic actuators are made of ceramic piezoelectric materials. While these materials are able to withstand large forces, they commonly will only deform a fraction of a percent. In the late 1990s, it has been demonstrated that some EAPs can exhibit up to a 380% strain, which is much more than any ceramic actuator. One of the most common applications for EAPs is in the field of robotics in the development of artificial muscles; thus, an electroactive polymer is often referred to as an artificial muscle.

History

The field of EAPs emerged back in 1880, when Wilhelm Röntgen designed an experiment in which he tested the effect of an electrostatic field on the mechanical properties of a stripe of natural rubber The rubber stripe was fixed at one end and was attached to a mass at the other. Electric charges were then sprayed onto the rubber and it was observed that the length changed by several centimeters. M.P. Sacerdote followed up on Roentgen's experiment by formulating a theory on strain response to an applied electric field in 1899. It wasn't until 1925 that the first piezoelectric polymer was discovered (Electret). Electret was formed by combining carnauba wax, rosin and beeswax, and then cooling the solution while it is subject to an applied DC electrical bias. The mixture would then solidify into a polymeric material that exhibited a piezoelectric effect.

Polymers that respond to environmental conditions other than an applied electric current have also been a large part of this area of study. In 1949 Katchalsky *et al.* demonstrated that when collagen filaments are dipped in acid or alkali solutions they would respond with a change in volume. The collagen filaments were found to expand in an acidic solution and contract in an alkali solution. Although other stimuli (such as pH) have been investigated, due to its ease and practicality most research has been devoted to developing polymers that respond to electrical stimuli in order to mimic biological systems.

The next major breakthrough in EAPs took place in the late 1960s. In 1969 Kawai demonstrated that polyvinylidene fluoride (PVDF) exhibits a large piezoelectric effect. This sparked research interest in developing other polymers systems that would show a similar effect. In 1977 the first electrically conducting polymers were discovered by Hideki Shirakawa *et al.* Shirakawa along with Alan MacDiarmid and Alan Heeger demonstrated that polyacetylene was electrically conductive, and that by doping it with iodine vapor, they could enhance its conductivity by 8 orders of magnitude. Thus the conductance was close to that of a metal. By the late 1980s a number of other polymers had been shown to exhibit a piezoelectric effect or were demonstrated to be conductive.

In the early 1990s ionic polymer-metal composites were developed and shown to exhibit electroactive properties far superior to previous EAPs. The major advantage of IPMCs was that they were able to show activation (deformation) at voltages as low as 1 or 2 volts. This is orders of magnitude less than any previous EAP. Not only was the activation energy for these materials much lower, but they could also undergo much larger deformations. IPMCs were shown to exhibit anywhere up to 380% strain, orders of magnitude larger than previously developed EAPs.

In 1999 Yoseph Bar-Cohen proposed the Armwrestling Match of EAP Robotic Arm Against Human Challenge. This was a challenge in which research groups around the world competed to design a robotic arm consisting of EAP muscles that could defeat a human in an arm wrestling match. The first challenge was held at the Electroactive Polymer Actuators and Devices Conference in 2005. Another major milestone of the field is that the first commercially developed device including EAPs as an artificial muscle was produced in 2002 by Eamex in Japan. This device was a fish that is able to swim on its own, moving its tail using an EAP muscle. But the progress in practical development is not satisfactory.

DARPA-funded research in the 1990s at SRI International and led by Ron Pelrine developed an electroactive polymer using silicone and acrylic polymers; the technology was spun off into the company Artificial Muscle in 2003, with industrial production beginning in 2008. In 2010, Artificial Muscle became a subsidiary of Bayer MaterialScience.

Types

EAP can have several configurations, but are generally divided in two principal classes: Dielectric and Ionic.

Dielectric

Dielectric EAPs are materials in which actuation is caused by electrostatic forces between two electrodes which squeeze the polymer. Dielectric elastomers are capable of very high strains and are fundamentally a capacitor that changes its capacitance when a voltage is applied by allowing the polymer to compress in thickness and expand in area due to the electric field. This type of EAP typically requires a large actuation voltage to produce high electric fields (hundreds to thousands of volts), but very low electrical power consumption. Dielectric EAPs require no power to keep the actuator at a given position. Examples are electrostrictive polymers and dielectric elastomers.

Ferroelectric Polymers

Ferroelectric polymers are a group of crystalline polar polymers that are also ferroelectric, meaning that they maintain a permanent electric polarization that can be reversed, or switched, in an external electric field. Ferroelectric polymers, such as polyvinylidene fluoride (PVDF), are used in acoustic transducers and electromechanical actuators because of their inherent piezoelectric response, and as heat sensors because of their inherent pyroelectric response.

Figure 1: Structure of Poly(vinylidene fluoride)

Electrostrictive Graft Polymers

Electrostrictive graft polymers consist of flexible backbone chains with branching side chains. The side chains on neighboring backbone polymers cross link and form crystal units. The backbone and side chain crystal units can then form polarized monomers, which contain atoms with partial charges and generate dipole moments, shown in Figure 2. When an electrical field is applied, a force is applied to each partial charge and causes rotation of the whole polymer unit. This rotation causes electrostrictive strain and deformation of the polymer.

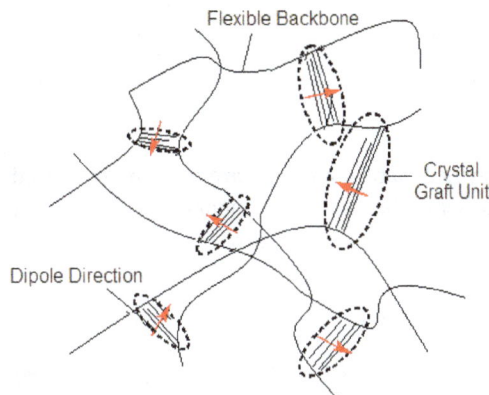

Figure 2: Cartoon of an electrostrictive graft polymer.

Liquid Crystalline Polymers

Main-chain liquid crystalline polymers have mesogenic groups linked to each other by a flexible spacer. The mesogens within a backbone form the mesophase structure causing the polymer itself to adopt a conformation compatible with the structure of the mesophase. The direct coupling of the liquid crystalline order with the polymer conformation has given main-chain liquid crystalline elastomers a large amount of interest. The synthesis of highly oriented elastomers leads to have a large strain thermal actuation along the polymer chain direction with temperature variation resulting in unique mechanical properties and potential applications as mechanical actuators.

Ionic

- Ionic EAPs, in which actuation is caused by the displacement of ions inside the polymer. Only a few volts are needed for actuation, but the ionic flow implies a higher electrical power needed for actuation, and energy is needed to keep the actuator at a given position. Examples of ionic EAPS are conductive polymers, ionic polymer-metal composites (IPMCs), and responsive gels. Yet another example is a Bucky gel actuator, which is a polymer-supported layer of polyelectrolyte material consisting of an ionic liquid sandwiched between two electrode layers consisting of a gel of ionic liquid containing single-wall carbon nanotubes. The name comes from the similarity of the gel to the paper that can be made by filtering carbon nanotubes, the so-called buckypaper.

Electrorheological Fluid

Electrorheological fluids change the viscosity of a solution with the application of an electric field. The fluid is a suspension of polymers in a low dielectric-constant liquid. With the application of a large electric field the viscosity of the suspension increases. Potential applications of these fluids include shock absorbers, engine mounts and acoustic dampers.

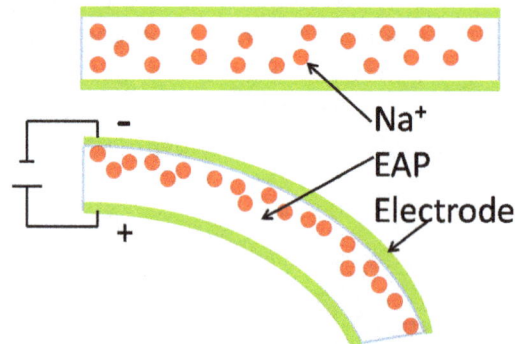

Figure 3: The cations in the ionic polymer-metal composite are randomly oriented in the absence of an electric field. Once a field is applied the cations gather to the side of the polymer in contact with the anode causing the polymer to bend.

Ionic Polymer-metal Composite

Ionic polymer-metal composites consist of a thin ionomeric membrane with noble metal electrodes plated on its surface. It also has cations to balance the charge of the anions fixed to the polymer backbone. They are very active actuators that show very high deformation at low applied voltage and show low impedance. Ionic polymer-metal composites work through electrostatic attraction between the cationic counter ions and the cathode of the applied electric field, a schematic representation is shown in Figure 3. These types of polymers show the greatest promise for bio-mimetic uses as collagen fibers are essentially composed of natural charged ionic polymers. Nafion and Flemion are commonly used ionic polymer metal composites.

Stimuli-responsive Gels

Stimuli-responsive gels (hydrogels, when the swelling agent is an aqueous solution) are a special kind of swellable polymer networks with volume phase transition behaviour. These materials change reversibly their volume, optical, mechanical and other properties by very small alterations of certain physical (e.g. electric field, light, temperature) or chemical (concentrations) stimuli. The volume change of these materials occurs by swelling/shrinking and is diffusion-based. Gels provide the biggest change in volume of solid-state materials. Combined with an excellent compatibility with micro fabrication technologies, especially stimuli-responsive hydrogels are of strong increasing interest for microsystems with sensors and actuators. Current fields of research and application are chemical sensor systems, microfluidics and multi-modal imaging systems.

Comparison of Dielectric and Ionic Eaps

Dielectric polymers are able to hold their induced displacement while activated under a DC voltage. This allows dielectric polymers to be considered for robotic applications. These types of materials also have high mechanical energy density and can be operated in air without a major decrease in performance. However, dielectric polymers require very high activation fields (>10 V/μm) that are close to the breakdown level.

The activation of ionic polymers, on the other hand, requires only 1-2 volts. They however need to maintain wetness, though some polymers have been developed as self-contained encapsulated activators which allows their use in dry environments. Ionic polymers also have a low electromechanical coupling. They are however ideal for bio-mimetic devices.

Characterization

While there are many different ways electroactive polymers can be characterized, only three will be addressed here: stress–strain curve, dynamic mechanical thermal analysis, and dielectric thermal analysis.

Stress-strain Curve

Stress strain curves provide information about the polymer's mechanical properties such as the brittleness, elasticity and yield strength of the polymer. This is done by providing a force to the polymer at a uniform rate and measuring the deformation that results. An example of this deformation is shown in Figure 4. This technique is useful for determining the type of material (brittle, tough, etc.), but it is a destructive technique as the stress is increased until the polymer fractures.

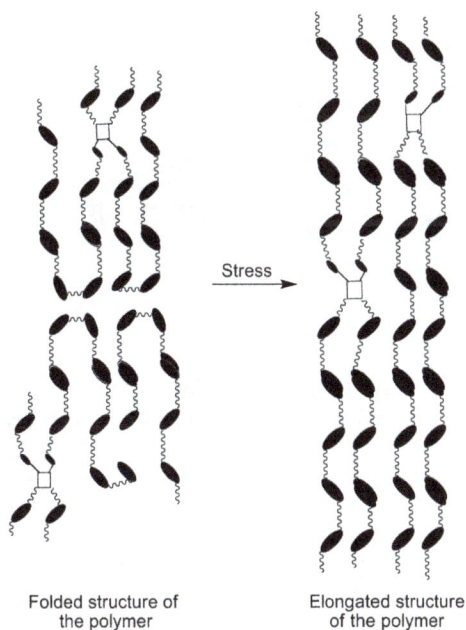

Figure 4: The unstressed polymer spontaneously forms a folded structure, upon application of a stress the polymer regains its original length.

Dynamic Mechanical Thermal Analysis (DMTA)

Both dynamic mechanical analysis is a non destructive technique that is useful in understanding the mechanism of deformation at a molecular level. In DMTA a sinusoidal stress is applied to the polymer, and based on the polymer's deformation the elastic modulus and damping characteristics are obtained (assuming the polymer is a damped harmonic oscillator). Elastic materials take the mechanical energy of the stress and convert it into potential energy which can later be recovered. An ideal spring will use all the potential energy to regain its original shape (no damping), while a liquid will use all the potential energy to flow, never returning to its original position or shape (high damping). A viscoeleastic polymer will exhibit a combination of both types of behavior.

Dielectric Thermal Analysis (DETA)

DETA is similar to DMTA, but instead of an alternating mechanical force an alternating electric field is applied. The applied field can lead to polarization of the sample, and if the polymer contains groups that have permanent dipoles (as in Figure 2), they will align with the electrical field. The permittivity can be measured from the change in amplitude and resolved into dielectric storage and loss components. The electric displacement field can also be measured by following the current. Once the field is removed, the dipoles will relax back into a random orientation.

Applications

EAP materials can be easily manufactured into various shapes due to the ease in processing many polymeric materials, making them very versatile materials. One potential application for EAPs is that they can potentially be integrated into microelectromechanical systems (MEMS) to produce smart actuators.

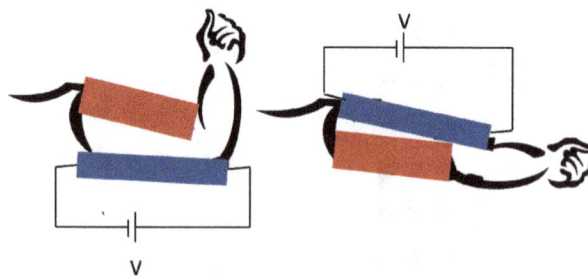

Figure 5: Cartoon drawing of an arm controlled by EAPs. When a voltage is applied (blue muscles) the polymer expands. When the voltage is removed (red muscles) the polymer returns to its original state.

Artificial Muscles

As the most prospective practical research direction, EAPs have been used in artificial muscles. Their ability to emulate the operation of biological muscles with high fracture toughness, large actuation strain and inherent vibration damping draw the attention of scientists in this field.

Tactile Displays

In recent years, "electro active polymers for refreshable Braille displays" has emerged to aid the visually impaired in fast reading and computer assisted communication. This concept is based on using an EAP actuator configured in an array form. Rows of electrodes on one side of an EAP film and columns on the other activate individual elements in the array. Each element is mounted with a Braille dot and is lowered by applying a voltage across the thickness of the selected element, causing local thickness reduction. Under computer control, dots would be activated to create tactile patterns of highs and lows representing the information to be read.

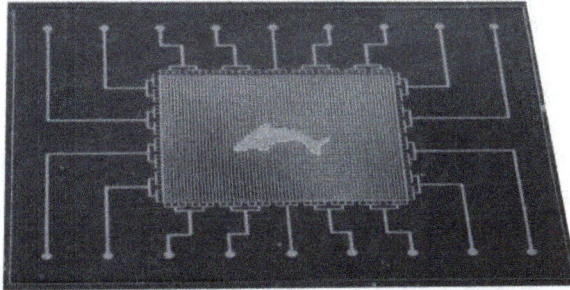

Figure 6: High resolution tactile display consisting of 4,320 (60x72) actuator pixels based on stimuli-responsive hydrogels. The integration density of the device is 297 components per cm². This display gives visual (monochromic) and physical (contours, relief, textures, softness) impressions of a virtual surface.

Visual and tactile impressions of a virtual surface are displayed by a high resolution tactile display, a so-called "artificial skin" (Fig.6) . These monolithic devices consist of an array of thousands of multimodal modulators (actuator pixels) based on stimuli-responsive hydrogels. Each modulator is able to change individually their transmission, height and softness. Besides their possible use as graphic displays for visually impaired such displays are interesting as free programmable keys of touchpads and consoles.

Microfluidics

EAP materials have huge potential for microfluidics e.g. as drug delivery systems, microfluidic devices and lab-on-a-chip. A first microfluidic platform technology reported in literature is based on stimuli-responsive gels. To avoid the electrolysis of water hydrogel-based microfluidic devices are mainly based on temperature-responsive polymers with lower critical solution temperature (LCST) characteristics, which are controlled by an electrothermic interface. Two types of micropumps are known, a diffusion micropump and a displacement micropump. Microvalves based on stimuli-responsive hydrogels show some advantageous properties such as particle tolerance, no leakage and outstanding pressure resistance. Besides these microfluidic standard components the hydrogel platform provides also chemical sensors and a novel class of microfluidic components, the chemical transistors (also referred as chemostat valves). These devices regulate a liquid flow if a threshold concentration of certain chemical is reached. Chemical transistors form the basis of microchemomechanical fluidic integrated circuits. "Chemical ICs" process exclusively chemical information, are energy-self-powered, operate automatically and are able for large-scale integration.

Another microfluidic platform is based on ionomeric materials. Pumps made from that material could offer low voltage (battery) operation, extremely low noise signature, high system efficiency, and highly accurate control of flow rate.

Another technology that can benefit from the unique properties of EAP actuators is optical membranes. Due to their low modulus, the mechanical impedance of the actuators, they are well-matched to common optical membrane materials. Also, a single EAP actuator is capable of generating displacements that range from micrometers to centimeters. For this reason, these materials can be used for static shape correction and jitter suppression. These actuators could also be used to correct for optical aberrations due to atmospheric interference.

Since these materials exhibit excellent electroactive character, EAP materials show potential in biomimetic-robot research, stress sensors and acoustics field, which will make EAPs become a more attractive study topic in the near future. They have been used for various actuators such as face muscles and arm muscles in humanoid robots.

Future Directions

The field of EAPs is far from mature, which leaves several issues that still need to be worked on. The performance and long-term stability of the EAP should be improved by designing a water impermeable surface. This will prevent the evaporation of water contained in the EAP, and also reduce the potential loss of the positive counter ions when the EAP is operating submerged in an aqueous environment. Improved surface conductivity should be explored using methods to produce a defect-free conductive surface. This could possibly be done using metal vapor deposition or other doping methods. It may also be possible to utilize conductive polymers to form a thick conductive layer. Heat resistant EAP would be desirable to allow operation at higher voltages without damaging the internal structure of the EAP due to the generation of heat in the EAP composite. Development of EAPs in different configurations (e.g., fibers and fiber bundles), would also be beneficial, in order to increase the range of possible modes of motion.

Roller Screw

A roller screw, also known as a planetary roller screw or satellite roller screw, is a low-friction precision screw-type actuator, a mechanical device for converting rotational motion to linear motion, or vice versa. Planetary roller screws are used as the actuating mechanism in many electro-mechanical linear actuators. Due to its complexity the roller screw is a relatively expensive actuator (as much as an order of magnitude more expensive than ball screws), but may be suitable for high-precision, high-speed, heavy-load, long-life and heavy-use applications.

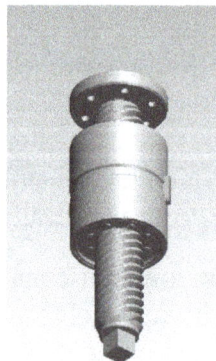

Roller screw mechanisms are commonly incorporated into motion/positioning systems in a variety of industries such as manufacturing and aerospace.

Principle of Operation

A roller screw is a mechanical actuator similar to a ball screw that uses rollers as the load transfer elements between nut and screw instead of balls. The rollers are typically threaded but may also be grooved depending on roller screw type. Providing more bearing points than ball screws within a given volume, roller screws can be more compact for a given load capacity while providing similar efficiency (75%-90%) at low to moderate speeds, and maintain relatively high efficiency at high speeds. Roller screws can surpass ball screws in regard to positioning precision, load rating, rigidity, speed, acceleration, and lifetime. Standard roller screw actuators can achieve dynamic load ratings above 130 tons of force (exceeded in single-unit actuator capacity only by hydraulic cylinders).

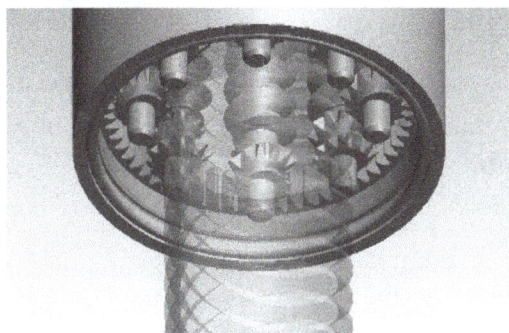

Standard roller screw timing

The three main elements of a typical planetary roller screw are the screw shaft, nut and planetary roller. The screw, a shaft with a multi-start V-shaped thread, provides a helical raceway for multiple rollers radially arrayed around the screw and encapsulated by a threaded nut. The thread of the screw is typically identical to the internal thread of the nut. The rollers spin in contact with, and serve as low-friction transmission elements between screw and nut. The rollers typically have a single-start thread with convex flanks that limit friction at the rollers' contacts with screw and nut. The rollers typically orbit the screw as they spin (in the manner of planet gears to sun gear), and are thus known as planetary, or satellite, rollers. As with a lead screw or ball screw, rotation of the nut results in screw travel, and rotation of the screw results in nut travel.

For a given screw diameter and quantity of thread starts more rollers corresponds to higher static load capacity, but not necessarily to a higher dynamic load capacity. Preloaded split nuts and double nuts are available to eliminate backlash.

Planetary Roller Screw Types

Carl Bruno Strandgren developed some of the earliest effective forms of roller screws and applied for a patent in Nice, France on February 1942. The French patent #888.281 was granted in August 1943 and published in December of the same year. The first commercial Roller Screw was designed and manufactured under his supervision in 1949 and was mounted on a narrow gauge locomotive which operated in a northern France coal mine. Subsequent units were produced and mounted on machine-tools and starting in 1955 on aircraft. At that time Carl Bruno Strandgren applied for a new patent incorporating detailed calculations and detailed manufacturing considerations for which he was awarded US patents for such a "Screw-Threaded Mechanism" in 1954, and "Nut and Screw Devices" and the "Roller Screw" in 1965.

Roller screw types are defined by the motion of the rollers relative to the nut and screw. The four commercially available types of roller screw are *standard, inverted, recirculating,* and *bearing ring*.

Differential roller screws, typically variants of the standard and recirculating types, are also commercially available. Differential roller screws modify the rotational speed ratios between the rollers and the screw by varying the flank angles and contact points of the threads or grooves. In that way differential roller screws change the effective lead of the screw. William J. Roantree received a US patent for the "Differential Roller Nut" in 1968.

Standard Planetary Roller Screw

The standard planetary roller screw is also known as the non-recirculating roller screw. The lack of axial movement of the roller relative to the nut, and the gearing of rollers to nut, are definitive of the standard type of roller screw.

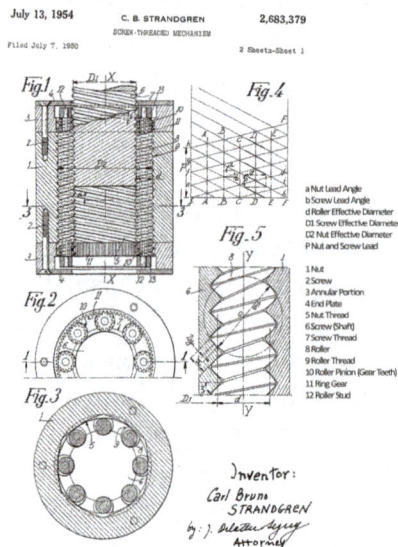

Patent drawing for standard roller screw (1954), with legend.

The nut and screw have identical multiple-start threads. The rollers have a single-start thread with an angle matching the nut thread. The matched thread angle prevents axial movement between the nut and the roller as the rollers spin. The nut assembly includes spacer rings and ring gears that position and guide the rollers. The spacer rings, which rotate within the ring gears, have equidistant holes that act as rotary bearings for the smooth pivot ends (studs) of the rollers. The ring gears time the spinning and orbit of the rollers about the screw axis by engaging gear teeth near the ends of the rollers. The spacer rings rotate on axis with the screw in unison with the orbit of the rollers. The spacer rings float relative to the nut, axially secured by retaining rings, because they spin around the screw at a lower frequency (angular velocity) than the nut.

Configuration

Standard roller screws are typically identified by screw diameter (typically ranging from 3.5mm – 200mm) and lead (1mm – 42mm). The threading of the screw (3 – 6 starts) is either rolled (lower capacity) or ground (higher capacity). The diameters of the nut and rollers (7 – 14 in quantity) are simple functions of the screw diameter and lead.

Where:

s_d is effective screw diameter, or pitch diameter

r_d is effective roller diameter

n_d is effective nut inside diameter

t is thread starts on nut and screw

l is screw lead

p is roller thread pitch

The following relationships apply to standard and inverted roller screws:

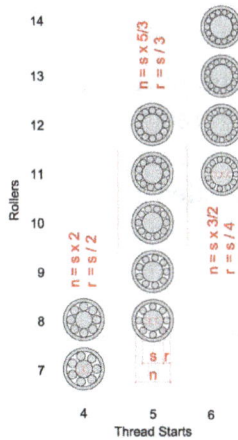

Common configurations of standard roller screws

nut to screw gear ratio

$$n_d = \frac{s_d t}{t-2} \qquad \therefore \qquad (n_d : s_d) = 1 : \frac{t}{t-2}$$

roller to screw gear ratio

$$r_d = \frac{s_d}{t-2} \qquad \therefore \qquad (r_d : s_d) = 1 : t-2$$

roller to nut gear ratio

$$r_d = \frac{n_d}{t} \qquad \therefore \qquad (r_d : n_d) = 1 : t$$

$$r_d = \frac{n_d - s_d}{2}$$

$$p = \frac{l}{t} \qquad \therefore \qquad \text{ratio of roller thread pitch to}$$

$$\text{screw lead } (p:l) = 1:t$$

For example, if

> Screw: 30 mm diameter, 20 mm lead, 5 start thread

then

> Rollers: 10 mm diameter rollers, 4 mm thread pitch

> Nut: 50 mm effective diameter.

Inverted Roller Screw

The inverted planetary roller screw is also known as the reverse roller screw. The lack of axial movement of the roller relative to the screw, and the gearing of rollers to screw, are definitive of the inverted type of planetary roller screw. This type of roller screw was developed simultaneously with the standard roller screw.

Inverted roller screws operate on the same principles of standard roller screws except that the function of the nut and screw is reversed in relation to the rollers. The rollers move axially within the nut, which is elongated to accommodate the full extent of screw shaft travel. The threaded portion of the screw shaft is limited to the threaded length of the rollers. The non-threaded portion of the screw shaft can be a smooth or non-cylindrical shape. The ring gear is replaced by gear teeth above and below the threaded portion of the screw shaft.

Aside from the inversion of the relationship of rollers to nut and screw, the configuration and relationships of inverted roller screws match those of standard roller screws.

Recirculating Roller Screw

The recirculating type of planetary roller screw is also known as a recycling roller screw. A recirculating roller screw can provide a very high degree of positional accuracy by using minimal thread leads. The rollers of a recirculating roller screw move axially within the nut until being reset after one orbit about the screw. Recirculating roller screws do not employ ring gears. Carl Bruno Strandgren was awarded a US Patent for the recirculating roller screw in 1965.

The screw and nut may have very fine identical single- or two-start threads. Recirculating rollers are grooved (instead of threaded) so they move axially during spinning engagement with the threads of the nut and screw, shifting up or down by one lead of thread after completing an orbit around the screw. The nut assembly typically includes a slotted cage and cam rings. The cage captivates the rollers in elongated slots, equally spacing the rollers while permitting rotation and axial motion of the rollers. The cam rings have opposing cams aligned with an axial groove in the wall of the nut. After a roller completes an orbit about the nut it is released into the groove, disengages from nut and screw, and is pushed between the cams to the axial midpoint of the nut assembly (shifting by a distance equal to the lead of the screw). Returned to its starting position, and reengaged to nut and screw, the roller may then orbit the screw once again.

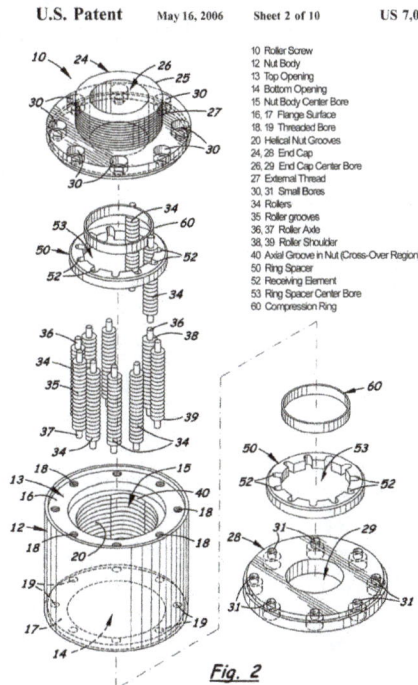

Cage-less recirculating roller screw patent drawing (2006), with legend.

In 2006, Charles C. Cornelius and Shawn P. Lawlor received a patent for a cage-less recirculating roller screw system. As with the traditional recirculating roller screw system, rollers disengage from the screw when they come upon an axial groove in the wall of the nut. The system differs in that the rollers are continually engaged by the nut, and the axial groove of the nut is threaded. Non-helical threads in the axial groove of the nut return the roller to its axial starting position (after completion of an orbit). Non-circular compression rings, or cam rings, at opposite ends of the rollers (roller axles) apply constant pressure between rollers and nut, synchronizing roller rotation and thrusting the rollers into the nut's axial groove. Lacking ring gears and roller cage, cage-less recirculating roller screws can be relatively efficient and, as a result, permit higher dynamic capacities for some screw shaft diameters.

Bearing Ring Roller Screw

In 1986 Oliver Saari was awarded a patent for a bearing ring roller screw, commonly referred to by its trademark, Spiracon. This type matches the orbit of the rollers to the rotation of the nut assembly. The actuator contains more load transfer elements than the other types, a bearing ring and thrust bearings, but manufacture of component parts is relatively simple (e.g. gearing teeth may be eliminated).

In the other roller screw types above, loads are transferred from the nut through the rollers to the screw (or in the reverse order). In this type of actuator, thrust bearings and a freely rotating internally grooved bearing ring transfer loads between the rollers and the nut.

The screw has a multi-start thread. The rollers and encapsulating rotating ring are identically grooved, not threaded, so there is no axial movement between the two. The nut assembly includes a cylindrical housing capped by non-rotating spacer rings. The spacer rings have equidistant holes that act as rotary bearings for the smooth pivot ends (studs) of the rollers. Roller-type thrust bearings between the spacer rings and bearing ring permit free rotation of the bearing ring while transferring the axial load between the two.

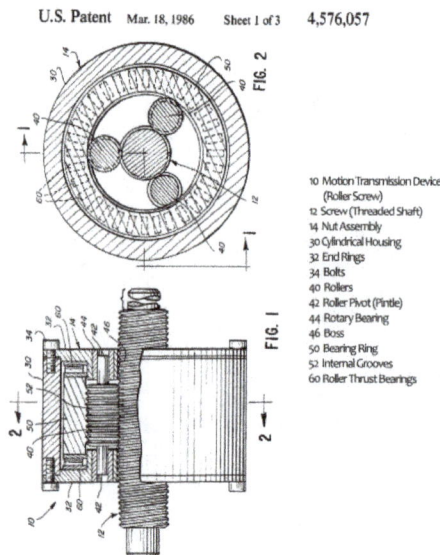

Patent Drawing for Spiracon Roller Screw (1986), with legend.

The rollers act as the "threads" of the nut assembly, causing axial movement of the rotating screw due to their orbital restraint. Screw rotation spins the rollers, which spin the bearing ring, dissipating the load-induced friction along the way.

Timothy A. Erhart was awarded a US patent in 1996 for a linear actuator effectively incorporating an inverted bearing ring roller screw. The screw shaft is grooved the length of and to match the grooved rollers, which travel with the shaft. The bearing ring is elongated and internally threaded for the length of screw shaft travel. The nut assembly housing and sealed end ring forms the exterior of the actuator assembly.

MEMS Magnetic Actuator

A MEMS magnetic actuator is a device that uses the microelectromechanical systems (MEMS) to convert an electric current into a mechanical output by employing the well-known Lorentz Force Equation or the theory of Magnetism.

Overview of MEMS

Micro-Electro-Mechanical System (MEMS) technology is a process technology in which mechanical and electro-mechanical devices or structures are constructed using special micro-fabrication techniques. These techniques include: bulk micro-machining, surface micro-machining, LIGA, wafer bonding, etc.

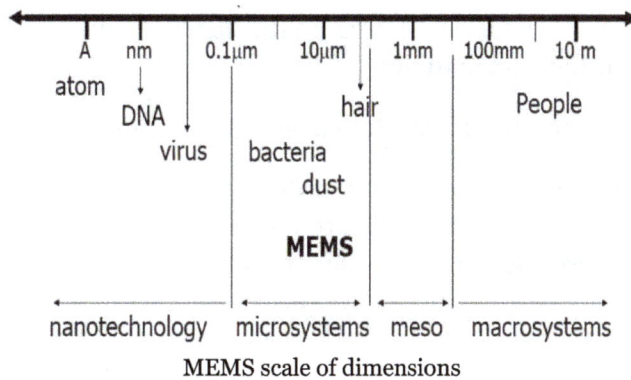

MEMS scale of dimensions

A device is considered to be a MEMS device if it satisfies the following:

- If its feature size is between 0.1 μm and hundreds of micrometers. (below this range, it becomes a nano device and above the range, it is considered a mesosystem)

- If it has some electrical functionality in its operation. This could include the generation of voltage by electromagnetic induction, by changing the gap between 2 electrodes or by a piezoelectric material.

- If the device has some mechanical functionality such as the deformation of a beam or diaphragm due to stress or strain.

- If it has a system-like functionality. The device must be integrable to other circuitries to form a system. This would be the interfacing circuitry and packaging for the device to become useful.

For the analysis of every MEMS device, the Lumped assumption is made: that if the size of the device is far less than the characteristic length scale of the phenomenon (wave or diffusion), then there would be no spatial variations across the entire device. Modelling becomes easy under this assumption.

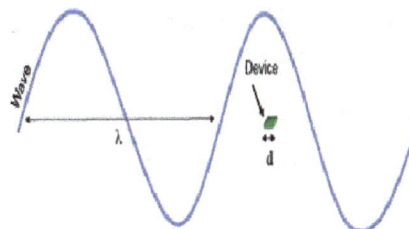

The Lumped Assumption

Operations in MEMS

The three major operations in MEMS are:

- Sensing: measuring a mechanical input by converting it to an electrical signal, e.g. a MEMS accelerometer or a pressure sensor (could also measure electrical signals as in the case of current sensors)

- Actuation: using an electrical signal to cause the displacement (or rotation) of a mechanical structure, e.g. a synthetic jet actuator.

- Power generation: generates power from a mechanical input, e.g. MEMS energy harvesters

These three operations require some form of transduction schemes, the most popular ones being: piezoelectric, electrostatic, piezoresistive, electrodynamic, magnetic and magnetostrictive. The MEMS magnetic actuators use the last three schemes for their operation.

Magnetic Actuation

The principle of magnetic actuation is based on the Lorentz Force Equation.

$$\vec{F}_{mag} = q\vec{v} \times B$$

When a current-carrying conductor is placed in a static magnetic field, the field produced around the conductor interacts with the static field to produce a force. This force can be used to cause the displacement of a mechanical structure.

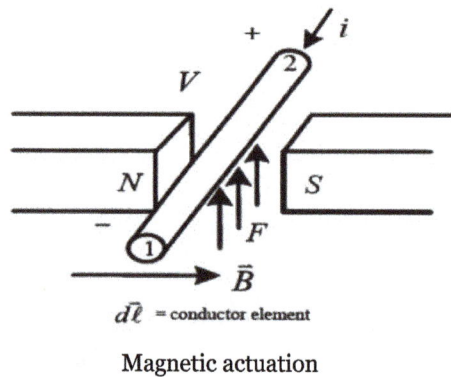

$d\vec{\ell}$ = conductor element

Magnetic actuation

Governing Equations and Parameters

A typical MEMS actuator is shown on the right. For a single turn of circular coil, the equations that govern its operation are:

- The H-field from a circular conductor:

$$H(z) = \frac{Ir^2}{2(r^2 + z^2)^{3/2}}$$

- The force produced by the interaction of the flux densities:

$$F_z = B_1 A_{mag} \int\limits_{z}^{z+h_{mag}} \frac{dHz}{dz} dz$$

The deflection of a mechanical structure for actuation depends on certain parameters of the device. For actuation, there has to be an applied force and a restoring force. The applied force is the force represented by the equation above, while the restoring force is fixed by the spring constant of the moving structure.

The applied force depends on both the field from the coils and the magnet. The remanence value of the magnet, its volume and position from the coils all contribute to its effect on the applied Force. Whereas the number of turns of coil, its size (radius) and the amount of current passing through it determines its effect on the Applied Force. The spring constant depends on the Young's Modulus of the moving structure, and its length, width and thickness.

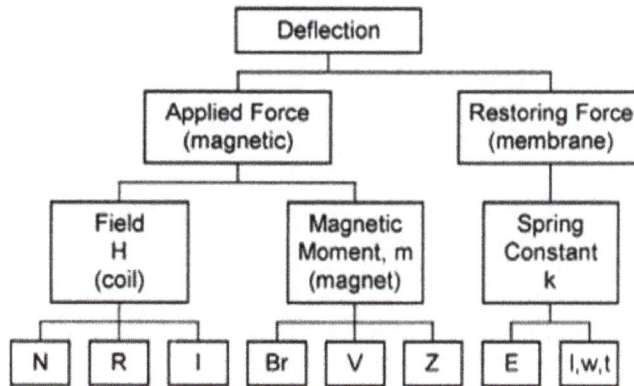

Magnetostrictive actuators

Magnetic actuation is not limited to the use of Lorentz force to cause a mechanical displacement. Magnetostrictive actuators can also use the theory of magnetism to bring about displacement. Materials that change their shapes when exposed to magnetic fields can now be used to drive high-reliability linear motors and actuators..An example is a nickel rod that tends to deform when it is placed in an external magnetic field. Another example is wrapping a series of electromagnetic induction coils around a metal tube in which a Terfenol-D material is placed. The coils generate a moving magnetic field that courses wavelike down the successive windings along the stator tube. As the traveling magnetic field causes each succeeding cross section of Terfenol-D to elongate, then contract when the field is removed, the rod will actually "crawl" down the stator tube like an inchworm. Repeated propagating waves of magnetic flux will translate the rod down the tube's length, producing a useful stroke and force output. The amount of motion generated by the material is proportional to the magnetic field provided by the coil system, which is a function of the electric current. This type of motive device, which features a single moving part, is called an elastic-wave or peristaltic linear motor. (view: Video of a Magnetostrictive micro walker)

Advantages of Magnetic Actuators

- High actuation force and stroke (displacement)

- Direct, fully linear transduction (in the case of electrodynamic actuation)

- Bi-directional actuation

- Contactless remote actuation

- Low-voltage actuation

- A figure of merit for actuators is the density of field energy that can be stored in the gap between the rotor and stator. Magnetic actuation has a potentially high energy density

Magnet Material

The operation of the magnetic actuator depends on the interaction between the field from an electromagnet and a static field. To produce this static field, it is important to use the right material. In MEMS, permanent magnets have become the favorite because they have a very good scaling factor and they retain their magnetization even when there is no external field... meaning that they need not be continuously magnetized when they are in use

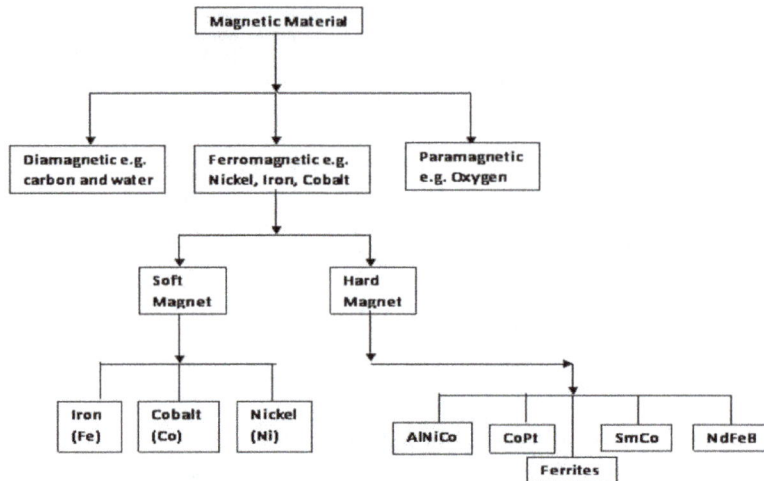

Magnet material selection for static B-Field

Integrating The Magnet into The Mems Device

As earlier discussed, MEMS devices are designed and fabricated using special micro-fabrication techniques. The major challenge however for magnetic MEMS is the integration of the magnet into the MEMS device. Recent research has suggested solutions to this challenge.

Fabrication (Or Molding) of The Magnet

There are several ways by which the magnet could be fabricated on a MEMS structure:

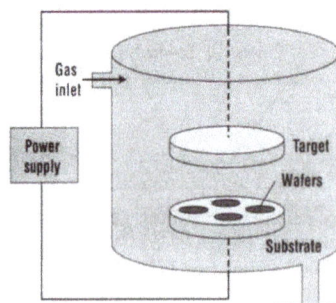

Sputtering

- Sputtering: Argon ion bombardment of the material release particles of the material. Mainly for depositing rare earth magnets. Deposition rate and film surface area depend on sputtering tool and target size

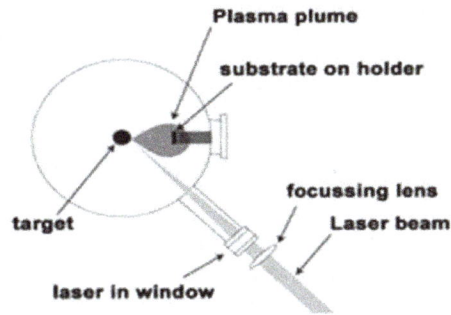

Pulsed Layer Deposition

- Pulsed Layer Deposition: a high-power pulsed laser beam is focused inside a vacuum chamber to strike a target of the material that is to be deposited

- Electroplating

- Screen printing

- Wax/parylene bonding

Issues with Magnetic Actuation

- High-power dissipation. This is a major problem for magnetic MEMS, but work is underway to circumvent this.

- Fabrication of the coil

- Integration of the micromagnet into the MEMS device

- Process-material compatibility

- Integratability into the overall microfabrication process (maintain cost and throughput)

- So that preexisting processes in the fabrication of the MEMS device will not be tampered with, deposition temperatures and post-deposition treatment/conditions must be tolerable. Also, the micromagnet must be able to withstand any chemical treatment that will come after its deposition

- Issues with magnetization (One may want to have more than one direction of magnetization; this creates a problem)

Each of these challenges can be mitigated or lessened by the right choice of material, choice of molding or fabrication method, and the type of device that is to be constructed. Applications of the magnetic actuator include: the synthetic jet actuator, micro-pumps and micro-relays.

References

- Parker, Dana T. Building Victory: Aircraft Manufacturing in the Los Angeles Area in World War II, p. 87, Cypress, CA, 2013. ISBN 978-0-9897906-0-4.

- Carlisle, Rodney (2004). Scientific American Inventions and Discoveries, p. 266. John Wiley & Sons, Inc., New Jersey. ISBN 0-471-24410-4.

- "Advantages of Hydraulic Presses". MetalFormingFacts.com. The Lubrizol Corporation. 4 February 2013. Retrieved 23 August 2013.

- De la Vergne, Jack (2003). Hard Rock Miner's Handbook. Tempe/North Bay: McIntosh Engineering. pp. 114–124. ISBN 0-9687006-1-6.

- Glass, J. Edward; Schulz, Donald N.; Zukosi, C.F (May 13, 1991). "1". Polymers as Rheology Modifiers. ACS Symposium Series. 462. Americal Chemical Society. pp. 2–17. ISBN 9780841220096.

- Cowie, J.M.G.; Valerai Arrighi (2008). "13". Polymers: Chemistry and Physics of Modern Material (Third ed.). Florida: CRC Press. pp. 363–373. ISBN 978-0-8493-9813-1.

- Kim, K.J.; Tadokoro, S. (2007). Electroactive Polymers for Robotic Applications, Artificial Muscles and Sensors. London: Springer. ISBN 978-1-84628-371-0.

Remote Sensing: An Overview

Remote sensing is the receiving of information about an object with no physical contact. The military, intelligence and economic planners usually practice it. This chapter is an overview of the subject matter incorporating all the major aspects of remote sensing. This chapter is a compilation of various ideas of remote sensing that form an integral part of the broader subject matter.

Water Remote Sensing

Water Remote Sensing studies the color of water through the observation of the spectrum of water leaving radiation. From the study of this spectrum, the concentration of optically active components of the upper layer of the water body can be concluded via specific algorithms. Water quality monitoring by remote sensing and close-range instruments has obtained considerable attention since the founding of EU Water Framework Directive.

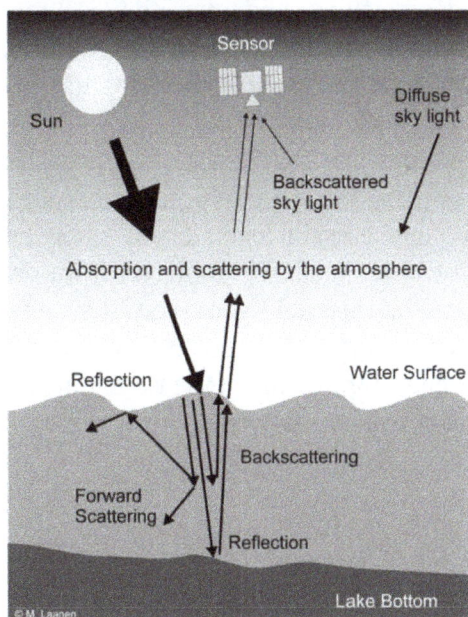

The path covered by light from the sun through the water body to the remote sensing sensor

History

If water remote sensing is defined as the observation of the water from a distance in order to describe its color, without taking water samples, the gradual development of understanding of the transparency of natural waters and of the reason of their clarity variability and coloration has been sketched from the times of Henry Hudson (1600) to those of Chandrasekhara Raman (1930). However, the development of water remote sensing techniques (by the use of satellite imaging, aircraft or close range optical devices) didn't start until the early 1970s. These first techniques measured the spectral and thermal differences in the emitted energy from water surfaces. In general, empirical relationships were settled between the spectral properties and the water quality parameters of the water body. In 1974, Ritchie et al. (1974) developed an empirical approach to determine suspended sediments. This kind of empirical models are only able to use to determine water quality parameters of water bodies with similar conditions. In 1992 an analytical approach was used by Schiebe et al. (1992). This approach was based on the optical characteristics of water and water quality parameters to elaborate a physically based model of the relationship between the spectral and physical properties of the surface water studied. This physically based model was successfully applied in order to estimate suspended sediment concentrations.

Function

Water Remote sensing instruments allow to record the color of a water body, which provides information on the presence and abundance of optically active natural water components. The water color spectrum is defined as an apparent optical property (AOP) of the water. This means that the color of the water is influenced by the angular distribution of the light field and by the nature and quantity of the substances in the medium, in this case, water. Thus, the value of this parameter will change with changes in the optical properties and concentrations of the optically active substances in the water, the inherent optical properties or IOPS. The IOPS are independent from the angular distribution of light but they are dependent from the type and substances present in the medium as well. For instance, the diffuse attenuation coefficient of downwelling irradiance, Kd (it is often used as an index of water clarity or ocean turbidity) is defined as an AOP, while the absorption coefficient and the scattering coefficient of the medium are defined as IOPS. There are two different approaches to determine the concentration of optically active water components by the study of the spectra. The first approach consist of empirical algorithms based on statistical relationships and the second approach consists of analytical algorithms based on the inversion of calibrated bio-optical models. Accurate calibration of the relationships/models used is an important condition for successful inversion on water remote sensing techniques and the determination of concentration of water quality parameters from observed spectral remote sensing data. Thus, these techniques depend on their ability to record these changes in the spectral signature of light backscattered from water surface and relate these recorded changes to water quality parameters via empirical or analytical approaches. Depending of the water constituents of interest and the sensor used, different parts of the spectrum will be analysed.

Contribution

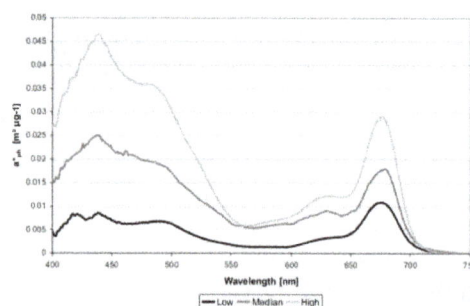

Example of specific phytoplankton absorption spectra. In this graph the characteristic blue and red Ch-a peaks at 438 nm and 676 nm can be seen. Another visible peak is the Cyanophicocianin absorption maximum at 624 nm.

By the use of Optical close range devices (e.g. spectrometers, radiometers), airplanes or helicopters (airborne remote sensing) and satellites (space born remote sensing), the light reflected from the water bodies is measured. For instance, algorithms are used to retrieve parameters such as chlorophyll-a(Chl-a) and Suspended Particulate Matter (SPM) concentration, the absorption by colored dissolved organic matter at 440 nm (aCDOM) and secchi depth. The measurement of these values will give an idea about the water quality of the water body being studied. A very high concentration of green pigments like chlorophyll might indicate the presence of an algal bloom, for example, due to eutrophication processes. Thus, the chlorophyll concentration could be used as a proxy or indicator for the trophic condition of a water body. In the same manner, other optical quality parameters such as suspended particles or Suspended Particulate matter (SPM), Colored Dissolved Organic Matter (CDOM), Transparency (Kd), and chlorophyll-a (Chl-a) can be used to monitor water quality.

Remote Sensing

Remote sensing is the acquisition of information about an object or phenomenon without making physical contact with the object and thus in contrast to on site observation. Remote sensing is used in numerous fields, including geography and most Earth Science disciplines (for example, hydrology, ecology, oceanography, glaciology, geology); it also has military, intelligence, commercial, economic, planning, and humanitarian applications. In modern usage, the term generally refers to the use of aerial sensor technologies to detect and classify objects on Earth (both on the surface, and in the atmosphere and oceans) by means of propagated signals (e.g. electromagnetic radiation). It may be split into active remote sensing (when a signal is first emitted from aircraft or satellites) or passive (e.g. sunlight) when information is merely recorded.

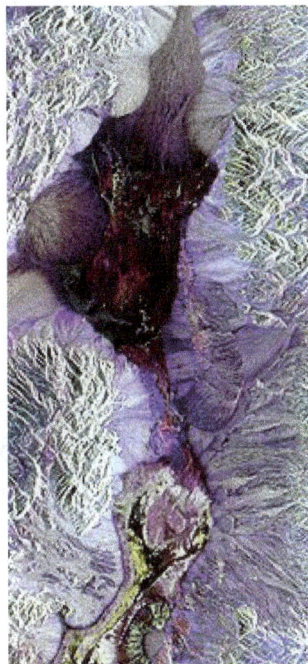

Synthetic aperture radar image of Death Valley colored using polarimetry.

Overview

Passive sensors gather radiation that is emitted or reflected by the object or surrounding areas. Reflected sunlight is the most common source of radiation measured by passive sensors. Examples of passive remote sensors include film photography, infrared, charge-coupled devices, and radiometers. Active collection, on the other hand, emits energy in order to scan objects and areas whereupon a sensor then detects and measures the radiation that is reflected or backscattered from the target. RADAR and LiDAR are examples of active remote sensing where the time delay between emission and return is measured, establishing the location, speed and direction of an object.

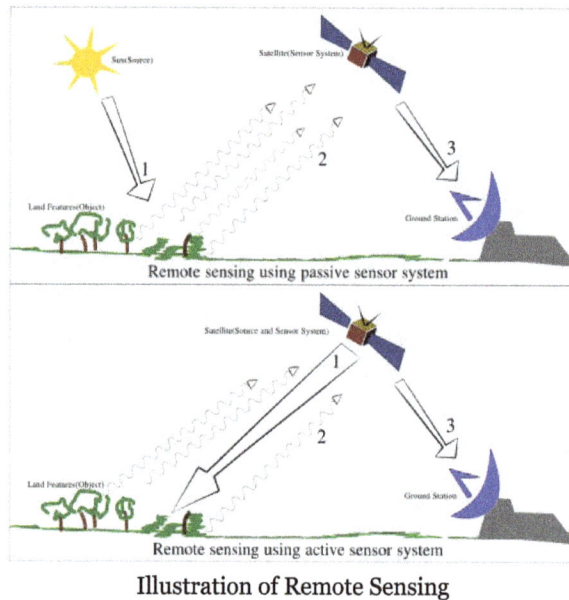

Illustration of Remote Sensing

Remote sensing makes it possible to collect data of dangerous or inaccessible areas. Remote sensing applications include monitoring deforestation in areas such as the Amazon Basin, glacial features in Arctic and Antarctic regions, and depth sounding of coastal and ocean depths. Military collection during the Cold War made use of stand-off collection of data about dangerous border areas. Remote sensing also replaces costly and slow data collection on the ground, ensuring in the process that areas or objects are not disturbed.

Orbital platforms collect and transmit data from different parts of the electromagnetic spectrum, which in conjunction with larger scale aerial or ground-based sensing and analysis, provides researchers with enough information to monitor trends such as El Niño and other natural long and short term phenomena. Other uses include different areas of the earth sciences such as natural resource management, agricultural fields such as land usage and conservation, and national security and overhead, ground-based and stand-off collection on border areas.

Data Acquisition Techniques

The basis for multispectral collection and analysis is that of examined areas or objects that reflect or emit radiation that stand out from surrounding areas. For a summary of major remote sensing satellite systems.

Applications of Remote Sensing

- Conventional radar is mostly associated with aerial traffic control, early warning, and certain large scale meteorological data. Doppler radar is used by local law enforcements' monitoring of speed limits and in enhanced meteorological collection such as wind speed and direction within weather systems in addition to precipitation location and intensity. Other types of active collection includes plasmas in the ionosphere. Interferometric synthetic aperture radar is used to produce precise digital elevation models of large scale terrain.

- Laser and radar altimeters on satellites have provided a wide range of data. By measuring the bulges of water caused by gravity, they map features on the seafloor to a resolution of a mile or so. By measuring the height and wavelength of ocean waves, the altimeters measure wind speeds and direction, and surface ocean currents and directions.

- Ultrasound (acoustic) and radar tide gauges measure sea level, tides and wave direction in coastal and offshore tide gauges.

- Light detection and ranging (LIDAR) is well known in examples of weapon ranging, laser illuminated homing of projectiles. LIDAR is used to detect and measure the concentration of various chemicals in the atmosphere, while airborne LIDAR can be used to measure heights of objects and features on the ground more accurately than with radar technology. Vegetation remote sensing is a principal application of LIDAR.

- Radiometers and photometers are the most common instrument in use, collecting reflected and emitted radiation in a wide range of frequencies. The most common are visible and infrared sensors, followed by microwave, gamma ray and rarely, ultraviolet. They may also be used to detect the emission spectra of various chemicals, providing data on chemical concentrations in the atmosphere.

- Stereographic pairs of aerial photographs have often been used to make topographic maps by imagery and terrain analysts in trafficability and highway departments for potential routes, in addition to modelling terrestrial habitat features.

- Simultaneous multi-spectral platforms such as Landsat have been in use since the 70's. These thematic mappers take images in multiple wavelengths of electro-magnetic radiation (multi-spectral) and are usually found on Earth observation satellites, including (for example) the Landsat program or the IKONOS satellite. Maps of land cover and land use from thematic mapping can be used to prospect for minerals, detect or monitor land usage, detect invasive vegetation, deforestation, and examine the health of indigenous plants and crops, including entire farming regions or forests. Landsat images are used by regulatory agencies such as KY-DOW to indicate water quality parameters including Secchi depth, chlorophyll a density and total phosphorus content. Weather satellites are used in meteorology and climatology.

- Hyperspectral imaging produces an image where each pixel has full spectral information with imaging narrow spectral bands over a contiguous spectral range. Hyperspectral imagers are used in various applications including mineralogy, biology, defence, and environmental measurements.

- Within the scope of the combat against desertification, remote sensing allows to follow-up and monitor risk areas in the long term, to determine desertification factors, to support decision-makers in defining relevant measures of environmental management, and to assess their impacts.

Geodetic

- Overhead geodetic collection was first used in aerial submarine detection and gravitational data used in military maps. This data revealed minute perturbations in the Earth's gravitational field (geodesy) that may be used to determine changes in the mass distribution of the Earth, which in turn may be used for geological studies.

Acoustic and Near-Acoustic

- Sonar: *passive sonar*, listening for the sound made by another object (a vessel, a whale etc.); *active sonar*, emitting pulses of sounds and listening for echoes, used for detecting, ranging and measurements of underwater objects and terrain.

- Seismograms taken at different locations can locate and measure earthquakes (after they occur) by comparing the relative intensity and precise timings.

- Ultrasound: Ultrasound sensors, that emit high frequency pulses and listening for echoes, used for detecting water waves and water level, as in tide gauges or for towing tanks.

To coordinate a series of large-scale observations, most sensing systems depend on the following: platform location and the orientation of the sensor. High-end instruments now often use positional information from satellite navigation systems. The rotation and orientation is often provided within a degree or two with electronic compasses. Compasses can measure not just azimuth (i. e. degrees to magnetic north), but also altitude (degrees above the horizon), since the magnetic field curves into the Earth at different angles at different latitudes. More exact orientations require gyroscopic-aided orientation, periodically realigned by different methods including navigation from stars or known benchmarks.

Data Processing

Generally speaking, remote sensing works on the principle of the *inverse problem*. While the object or phenomenon of interest (the state) may not be directly measured, there exists some other variable that can be detected and measured (the observation) which may be related to the object of interest through a calculation. The common analogy given to describe this is trying to determine the type of animal from its footprints. For example, while it is impossible to directly measure temperatures in the upper atmosphere, it is possible to measure the spectral emissions from a known chemical species (such as carbon dioxide) in that region. The frequency of the emissions may then be related via thermodynamics to the temperature in that region.

The quality of remote sensing data consists of its spatial, spectral, radiometric and temporal resolutions.

Spatial resolution

The size of a pixel that is recorded in a raster image – typically pixels may correspond to square areas ranging in side length from 1 to 1,000 metres (3.3 to 3,280.8 ft).

Spectral resolution

The wavelength width of the different frequency bands recorded – usually, this is related to the number of frequency bands recorded by the platform. Current Landsat collection is that of seven bands, including several in the infra-red spectrum, ranging from a spectral resolution of 0.07 to 2.1 µm. The Hyperion sensor on Earth Observing-1 resolves 220 bands from 0.4 to 2.5 µm, with a spectral resolution of 0.10 to 0.11 µm per band.

Radiometric resolution

The number of different intensities of radiation the sensor is able to distinguish. Typically, this ranges from 8 to 14 bits, corresponding to 256 levels of the gray scale and up to 16,384 intensities or "shades" of colour, in each band. It also depends on the instrument noise.

Temporal resolution

The frequency of flyovers by the satellite or plane, and is only relevant in time-series studies or those requiring an averaged or mosaic image as in deforesting monitoring. This was first used by the intelligence community where repeated coverage revealed changes in infra-structure, the deployment of units or the modification/introduction of equipment. Cloud cover over a given area or object makes it necessary to repeat the collection of said location.

In order to create sensor-based maps, most remote sensing systems expect to extrapolate sensor data in relation to a reference point including distances between known points on the ground. This depends on the type of sensor used. For example, in conventional photographs, distances are accurate in the center of the image, with the distortion of measurements increasing the farther you get from the center. Another factor is that of the platen against which the film is pressed can cause severe errors when photographs are used to measure ground distances. The step in which this problem is resolved is called georeferencing, and involves computer-aided matching of points in the image (typically 30 or more points per image) which is extrapolated with the use of an es-tablished benchmark, "warping" the image to produce accurate spatial data. As of the early 1990s, most satellite images are sold fully georeferenced.

In addition, images may need to be radiometrically and atmospherically corrected.

Radiometric correction

Allows to avoid radiometric errors and distortions. The illumination of objects on the Earth surface is uneven because of different properties of the relief. This factor is taken into ac-count in the method of radiometric distortion correction. Radiometric correction gives a scale to the pixel values, e. g. the monochromatic scale of 0 to 255 will be converted to actual radiance values.

Topographic correction (also called terrain correction)

In rugged mountains, as a result of terrain, the effective illumination of pixels varies considerably. In a remote sensing image, the pixel on the shady slope receives weak illumination and has a low radiance value, in contrast, the pixel on the sunny slope receives strong illumination and has a high radiance value. For the same object, the pixel radiance value on the shady slope will be different from that on the sunny slope. Additionally, different objects may have similar radiance values. These ambiguities seriously affected remote sensing image information extraction accuracy in mountainous areas. It became the main obstacle to further application of remote sensing images. The purpose of topographic correction is to eliminate this effect, recovering the true reflectivity or radiance of objects in horizontal conditions. It is the premise of quantitative remote sensing application.

Atmospheric correction

Elimination of atmospheric haze by rescaling each frequency band so that its minimum value (usually realised in water bodies) corresponds to a pixel value of 0. The digitizing of data also makes it possible to manipulate the data by changing gray-scale values.

Interpretation is the critical process of making sense of the data. The first application was that of aerial photographic collection which used the following process; spatial measurement through the use of a light table in both conventional single or stereographic coverage, added skills such as the use of photogrammetry, the use of photomosaics, repeat coverage, Making use of objects' known dimensions in order to detect modifications. Image Analysis is the recently developed automated computer-aided application which is in increasing use.

Object-Based Image Analysis (OBIA) is a sub-discipline of GIScience devoted to partitioning remote sensing (RS) imagery into meaningful image-objects, and assessing their characteristics through spatial, spectral and temporal scale.

Old data from remote sensing is often valuable because it may provide the only long-term data for a large extent of geography. At the same time, the data is often complex to interpret, and bulky to store. Modern systems tend to store the data digitally, often with lossless compression. The difficulty with this approach is that the data is fragile, the format may be archaic, and the data may be easy to falsify. One of the best systems for archiving data series is as computer-generated machine-readable ultrafiche, usually in typefonts such as OCR-B, or as digitized half-tone images. Ultrafiches survive well in standard libraries, with lifetimes of several centuries. They can be created, copied, filed and retrieved by automated systems. They are about as compact as archival magnetic media, and yet can be read by human beings with minimal, standardized equipment.

Data Processing Levels

To facilitate the discussion of data processing in practice, several processing "levels" were first defined in 1986 by NASA as part of its Earth Observing System and steadily adopted since then, both internally at NASA (e. g.,) and elsewhere (e. g.,); these definitions are:

Level	Description
0	Reconstructed, unprocessed instrument and payload data at full resolution, with any and all communications artifacts (e. g., synchronization frames, communications headers, duplicate data) removed.

1a	Reconstructed, unprocessed instrument data at full resolution, time-referenced, and annotated with ancillary information, including radiometric and geometric calibration coefficients and georeferencing parameters (e. g., platform ephemeris) computed and appended but not applied to the Level 0 data (or if applied, in a manner that level 0 is fully recoverable from level 1a data).
1b	Level 1a data that have been processed to sensor units (e. g., radar backscatter cross section, brightness temperature, etc.); not all instruments have Level 1b data; level 0 data is not recoverable from level 1b data.
2	Derived geophysical variables (e. g., ocean wave height, soil moisture, ice concentration) at the same resolution and location as Level 1 source data.
3	Variables mapped on uniform spacetime grid scales, usually with some completeness and consistency (e. g., missing points interpolated, complete regions mosaicked together from multiple orbits, etc.).
4	Model output or results from analyses of lower level data (i. e., variables that were not measured by the instruments but instead are derived from these measurements).

A Level 1 data record is the most fundamental (i. e., highest reversible level) data record that has significant scientific utility, and is the foundation upon which all subsequent data sets are produced. Level 2 is the first level that is directly usable for most scientific applications; its value is much greater than the lower levels. Level 2 data sets tend to be less voluminous than Level 1 data because they have been reduced temporally, spatially, or spectrally. Level 3 data sets are generally smaller than lower level data sets and thus can be dealt with without incurring a great deal of data handling overhead. These data tend to be generally more useful for many applications. The regular spatial and temporal organization of Level 3 datasets makes it feasible to readily combine data from different sources.

While these processing levels are particularly suitable for typical satellite data processing pipelines, other data level vocabularies have been defined and may be appropriate for more heterogeneous workflows.

History

The modern discipline of remote sensing arose with the development of flight. The balloonist G. Tournachon (alias Nadar) made photographs of Paris from his balloon in 1858. Messenger pigeons, kites, rockets and unmanned balloons were also used for early images. With the exception of balloons, these first, individual images were not particularly useful for map making or for scientific purposes.

The TR-1 reconnaissance/surveillance aircraft.

Systematic aerial photography was developed for military surveillance and reconnaissance purposes beginning in World War I and reaching a climax during the Cold War with the use of modified combat aircraft such as the P-51, P-38, RB-66 and the F-4C, or specifically designed collection platforms such as the U2/TR-1, SR-71, A-5 and the OV-1 series both in overhead and stand-off collection. A more recent development is that of increasingly smaller sensor pods such as those used by law enforcement and the military, in both manned and unmanned platforms. The advantage of this approach is that this requires minimal modification to a given airframe. Later imaging technologies would include Infra-red, conventional, Doppler and synthetic aperture radar.

The *2001 Mars Odyssey* used spectrometers and imagers to hunt for evidence of past or present water and volcanic activity on Mars.

The development of artificial satellites in the latter half of the 20th century allowed remote sensing to progress to a global scale as of the end of the Cold War. Instrumentation aboard various Earth observing and weather satellites such as Landsat, the Nimbus and more recent missions such as RADARSAT and UARS provided global measurements of various data for civil, research, and military purposes. Space probes to other planets have also provided the opportunity to conduct remote sensing studies in extraterrestrial environments, synthetic aperture radar aboard the Magellan spacecraft provided detailed topographic maps of Venus, while instruments aboard SOHO allowed studies to be performed on the Sun and the solar wind, just to name a few examples.

Recent developments include, beginning in the 1960s and 1970s with the development of image processing of satellite imagery. Several research groups in Silicon Valley including NASA Ames Research Center, GTE, and ESL Inc. developed Fourier transform techniques leading to the first notable enhancement of imagery data. In 1999 the first commercial satellite (IKONOS) collecting very high resolution imagery was launched.

Training and Education

At most universities remote sensing is associated with Geography departments. Remote Sensing has a growing relevance in the modern information society. It represents a key technology as part of the aerospace industry and bears increasing economic relevance – new sensors e.g. TerraSAR-X and RapidEye are developed constantly and the demand for skilled labour is increasing steadily. Furthermore, remote sensing exceedingly influences everyday life, ranging from weather forecasts to reports on climate change or natural disasters. As an example, 80% of the German students use the services of Google Earth; in 2006 alone the software was downloaded 100 million times. But studies have shown that only a fraction of them know more about the data they are working with. There exists a huge knowledge gap between the application and the understanding of satellite images. Remote sensing only plays a tangential role in schools, regardless of the political claims to strengthen the support for teaching on the subject. A lot of the computer software explicitly developed for school lessons has not yet been implemented due to its complexity. Thereby, the subject is either not at all integrated into the curriculum or does not pass the step of an interpretation of analogue images. In fact, the subject of remote sensing requires a consolidation of physics and mathematics as well as competences in the fields of media and methods apart from the mere visual interpretation of satellite images.

Many teachers have great interest in the subject "remote sensing", being motivated to integrate this topic into teaching, provided that the curriculum is considered. In many cases, this encouragement fails because of confusing information. In order to integrate remote sensing in a sustainable manner organizations like the EGU or digital earth encourages the development of learning modules and learning portals (e.g. FIS – Remote Sensing in School Lessons or Landmap – Spatial Discovery) promoting media and method qualifications as well as independent working.

Remote Sensing Software

Remote sensing data are processed and analyzed with computer software, known as a remote sensing application. A large number of proprietary and open source applications exist to process remote sensing data. Remote sensing software packages include:

- PCI Geomatica made by PCI Geomatics,

- TacitView from 2d3

- Socet GXP from BAE Systems,

- TNTmips from MicroImages,

- IDRISI from Clark Labs,

- eCognition from Trimble,

- and RemoteView made by Overwatch Textron Systems.

- Dragon/ips is one of the oldest remote sensing packages still available, and is in some cases free.

- ERDAS IMAGINE from Hexagon Geospatial (Separated from Intergraph SG&I),

- ENVI/IDL from Exelis Visual Information Solutions,

Open source remote sensing software includes:

- Opticks (software),

- Orfeo toolbox

- Others mixing remote sensing and GIS capabilities are: GRASS GIS, ILWIS, QGIS, and TerraLook.

According to an NOAA Sponsored Research by Global Marketing Insights, Inc. the most used applications among Asian academic groups involved in remote sensing are as follows: ERDAS 36% (ERDAS IMAGINE 25% & ERMapper 11%); ESRI 30%; ITT Visual Information Solutions ENVI 17%; MapInfo 17%.

Among Western Academic respondents as follows: ESRI 39%, ERDAS IMAGINE 27%, MapInfo 9%, and AutoDesk 7%.

Lidar

Lidar (also called LIDAR, LiDAR, and LADAR) is a surveying method that measures distance to a target by illuminating that target with a laser light. The name *lidar*, sometimes considered an acronym of *Light Detection And Ranging*, (sometimes *Light Imaging, Detection, And Ranging*), was originally a portmanteau of *light* and *radar*. Lidar is popularly used to make high-resolution maps, with applications in geodesy, geomatics, archaeology, geography, geology, geomorphology, seismology, forestry, atmospheric physics, laser guidance, airborne laser swath mapping (ALSM), and laser altimetry. Lidar sometimes is called *laser scanning* and *3D scanning*, with terrestrial, airborne, and mobile applications.

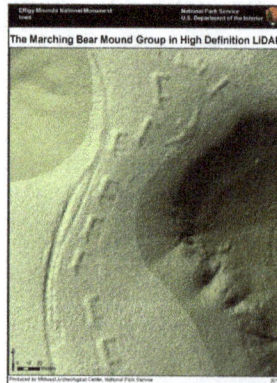

Lidar-derived image of Marching Bears Mound Group, Effigy Mounds National Monument.

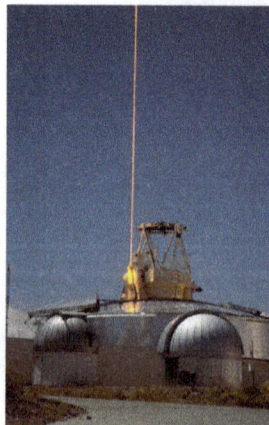

A FASOR used at the Starfire Optical Range for lidar and laser guide star experiments is tuned to the sodium D2a line and used to excite sodium atoms in the upper atmosphere.

This lidar may be used to scan buildings, rock formations, etc., to produce a 3D model. The lidar can aim its laser beam in a wide range: its head rotates horizontally; a mirror tilts vertically. The laser beam is used to measure the distance to the first object on its path.

History and Etymology

Lidar originated in the early 1960s, shortly after the invention of the laser, and combined laser-focused imaging with radar's ability to calculate distances by measuring the time for a signal to return. Its first applications came in meteorology, where the National Center for Atmospheric Research used it to measure clouds. The general public became aware of the accuracy and usefulness of lidar systems in 1971 during the Apollo 15 mission, when astronauts used a laser altimeter to map the surface of the moon.

Although some sources treat the word "lidar" as an acronym, the term originated as a portmanteau of "light" and "radar". The first published mention of lidar, in 1963, makes this clear: "Eventually the laser may provide an extremely sensitive detector of particular wavelengths from distant objects. Meanwhile, it is being used to study the moon by 'lidar' (light radar) …" The *Oxford English Dictionary* supports this etymology.

The interpretation of "lidar" as an acronym ("LIDAR") came later, beginning in 1970, based on the assumption that since the base term "radar" originally started as an acronym for "RAdio Detection And Ranging", "LIDAR" must stand for "LIght Detection And Ranging", or for "Laser Imaging, Detection and Ranging". Although the English language no longer treats "radar" as an acronym and printed texts universally present the word uncapitalized, the word "lidar" became capitalized as "LIDAR" in some publications beginning in the 1980s. Currently no consensus exists on capitalization, reflecting uncertainty about whether or not "lidar" is an acronym, and if it is an acronym, whether it should appear in lower case, like "radar". Various publications refer to lidar as "LIDAR", "LiDAR", "LIDaR", or "Lidar". The USGS uses both "LIDAR" and "lidar", sometimes in the same document; the *New York Times* uses both "lidar" and "Lidar".

General Description

Lidar uses ultraviolet, visible, or near infrared light to image objects. It can target a wide range of materials, including non-metallic objects, rocks, rain, chemical compounds, aerosols, clouds and even single molecules. A narrow laser-beam can map physical features with very high resolutions; for example, an aircraft can map terrain at 30 cm resolution or better.

Lidar has been used extensively for atmospheric research and meteorology. Lidar instruments fitted to aircraft and satellites carry out surveying and mapping – a recent example being the U.S. Geological Survey Experimental Advanced Airborne Research Lidar. NASA has identified lidar as a key technology for enabling autonomous precision safe landing of future robotic and crewed lunar-landing vehicles.

Wavelengths vary to suit the target: from about 10 micrometers to the UV (approximately 250 nm). Typically light is reflected via backscattering. Different types of scattering are used for different lidar applications: most commonly Rayleigh scattering, Mie scattering, Raman scattering, and fluorescence. Based on different kinds of backscattering, the lidar can be accordingly called Rayleigh Lidar, Mie Lidar, Raman Lidar, Na/Fe/K Fluorescence Lidar, and so on. Suitable combinations of wavelengths can allow for remote mapping of atmospheric contents by identifying wavelength-dependent changes in the intensity of the returned signal.

Design

In general there are two kinds of lidar detection schemes: "incoherent" or direct energy detection (which is principally an amplitude measurement) and coherent detection (which is best for Doppler, or phase sensitive measurements). Coherent systems generally use optical heterodyne detection, which, being more sensitive than direct detection, allows them to operate at a much lower power but at the expense of more complex transceiver requirements.

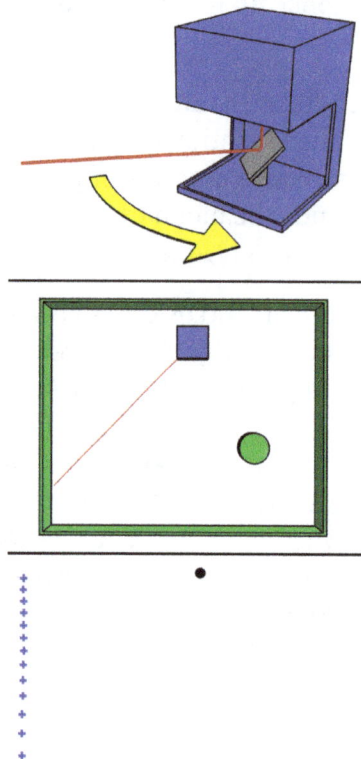

A basic lidar system involves a laser range finder reflected by a rotating mirror (top). The laser is scanned around the scene being digitised, in one or two dimensions (middle), gathering distance measurements at specified angle intervals (bottom).

In both coherent and incoherent lidar, there are two types of pulse models: *micropulse lidar* systems and *high energy* systems. Micropulse systems have developed as a result of the ever increasing amount of computer power available combined with advances in laser technology. They use considerably less energy in the laser, typically on the order of one microjoule, and are often "eye-safe," meaning they can be used without safety precautions. High-power systems are common in atmospheric research, where they are widely used for measuring many atmospheric parameters: the height, layering and densities of clouds, cloud particle properties (extinction coefficient, backscatter coefficient, depolarization), temperature, pressure, wind, humidity, trace gas concentration (ozone, methane, nitrous oxide, etc.).

There are several major components to a lidar system:

1. Laser — 600–1000 nm lasers are most common for non-scientific applications. They are inexpensive, but since they can be focused and easily absorbed by the eye, the maximum power is limited by the need to make them eye-safe. Eye-safety is often a requirement for most applications. A common alternative, 1550 nm lasers, are eye-safe at much higher power levels since this wavelength is not focused by the eye, but the detector technology is less advanced and so these wavelengths are generally used at longer ranges and lower accuracies. They are also used for military applications as 1550 nm is not visible in night vision goggles, unlike the shorter 1000 nm infrared laser. Airborne topographic mapping lidars generally use 1064 nm diode pumped YAG lasers, while bathymetric systems generally use 532 nm frequency doubled diode pumped YAG lasers because 532 nm penetrates water with much less attenuation than does 1064 nm. Laser settings include the laser repetition rate (which controls the data collection speed). Pulse length is generally an attribute of the laser cavity length, the number of passes required through the gain material (YAG, YLF, etc.), and Q-switch speed. Better target resolution is achieved with shorter pulses, provided the lidar receiver detectors and electronics have sufficient bandwidth.

2. Scanner and optics — How fast images can be developed is also affected by the speed at which they are scanned. There are several options to scan the azimuth and elevation, including dual oscillating plane mirrors, a combination with a polygon mirror, a dual axis scanner. Optic choices affect the angular resolution and range that can be detected. A hole mirror or a beam splitter are options to collect a return signal.

3. Photodetector and receiver electronics — Two main photodetector technologies are used in lidars: solid state photodetectors, such as silicon avalanche photodiodes, or photomultipliers. The sensitivity of the receiver is another parameter that has to be balanced in a lidar design.

4. Position and navigation systems — Lidar sensors that are mounted on mobile platforms such as airplanes or satellites require instrumentation to determine the absolute position and orientation of the sensor. Such devices generally include a Global Positioning System receiver and an Inertial Measurement Unit (IMU).

3D imaging can be achieved using both scanning and non-scanning systems. "3D gated viewing laser radar" is a non-scanning laser ranging system that applies a pulsed laser and a fast gated camera. Research has begun for virtual beam steering using DLP technology.

Imaging lidar can also be performed using arrays of high speed detectors and modulation sensitive detector arrays typically built on single chips using CMOS and hybrid CMOS/CCD fabrication techniques. In these devices each pixel performs some local processing such as demodulation or gating at high speed, downconverting the signals to video rate so that the array may be read like a camera. Using this technique many thousands of pixels / channels may be acquired simultaneously. High resolution 3D lidar cameras use homodyne detection with an electronic CCD or CMOS shutter.

A coherent imaging lidar uses synthetic array heterodyne detection to enable a staring single element receiver to act as though it were an imaging array.

In 2014 Lincoln Laboratory announced a new imaging chip with more than 16,384 pixels, each able to image a single photon, enabling them to capture a wide area in a single image. An earlier generation of the technology with one-quarter as many pixels was dispatched by the U.S. military after the January 2010 Haiti earthquake; a single pass by a business jet at 3,000 meters (10,000 ft.) over Port-au-Prince was able to capture instantaneous snapshots of 600-meter squares of the city at 30 centimetres (12 in), displaying the precise height of rubble strewn in city streets. The new system is another 10x faster. The chip uses indium gallium arsenide (InGaAs), which operates in the infrared spectrum at a relatively long wavelength that allows for higher power and longer ranges. In many applications, such as self-driving cars, the new system will lower costs by not requiring a mechanical component to aim the chip. InGaAs uses less hazardous wavelengths than conventional silicon detectors, which operate at visual wavelengths.

Types of Applications

Lidar has a wide range of applications which can be divided into airborne and terrestrial types. These different types of applications require scanners with varying specifications based on the data's purpose, the size of the area to be captured, the range of measurement desired, the cost of equipment, and more.

Airborne Lidar

Airborne lidar (also *airborne laser scanning*) is when a laser scanner, while attached to a plane during flight, creates a 3D point cloud model of the landscape. This is currently the most detailed and accurate method of creating digital elevation models, replacing photogrammetry. One major advantage in comparison with photogrammetry is the ability to filter out reflections from vegetation from the point cloud model to create a digital surface model which represents ground surfaces such as rivers, paths, cultural heritage sites, etc., which are concealed by trees. Within the category of airborne lidar, there is sometimes a distinction made between high-altitude and low-altitude applications, but the main difference is a reduction in both accuracy and point density of data acquired at higher altitudes. Airborne lidar can also be used to create bathymetric models in shallow water.

Drones are now being used with laser scanners, as well as other remote sensors, as a more economical method to scan smaller areas. The possibility of drone remote sensing also eliminates any danger that crews of a manned aircraft may be subjected to in difficult terrain or remote areas.

LIDAR scanning performed with the OnyxStar FOX-C8 HD UAV from AltiGator

Terrestrial Lidar

Terrestrial applications of lidar (also *terrestrial laser scanning*) happen on the Earth's surface and can be both stationary or mobile. Stationary terrestrial scanning is most common as a survey method, for example in conventional topography, monitoring, cultural heritage documentation and forensics. The 3D point clouds acquired from these types of scanners can be matched with digital images taken of the scanned area from the scanner's location to create realistic looking 3D models in a relatively short time when compared to other technologies. Each point in the point cloud is given the colour of the pixel from the image taken located at the same angle as the laser beam that created the point.

Mobile lidar (also *mobile laser scanning*) is when two or more scanners are attached to a moving vehicle to collect data along a path. These scanners are almost always paired with other kinds of equipment, including GNSS receivers and IMUs. One example application is surveying streets, where power lines, exact bridge heights, bordering trees, etc. all need to be taken into account. Instead of collecting each of these measurements individually in the field with a tachymeter, a 3D model from a point cloud can be created where all of the measurements needed can be made, depending on the quality of the data collected. This eliminates the problem of forgetting to take a measurement, so long as the model is available, reliable and has an appropriate level of accuracy.

Applications

This lidar-equipped mobile robot uses its lidar to construct a map and avoid obstacles.

There are a wide variety of applications for lidar, in addition to the applications listed below, as it is often mentioned in National lidar dataset programs.

Agriculture

Lidar also can be used to help farmers determine which areas of their fields to apply costly fertilizer. Lidar can create a topographical map of the fields and reveals the slopes and sun exposure of the farm land. Researchers at the Agricultural Research Service blended this topographical information with the farmland yield results from previous years. From this information, researchers categorized the farm land into high-, medium-, or low-yield zones. This technology is valuable to farmers because it indicates which areas to apply the expensive fertilizers to achieve the highest crop yield.

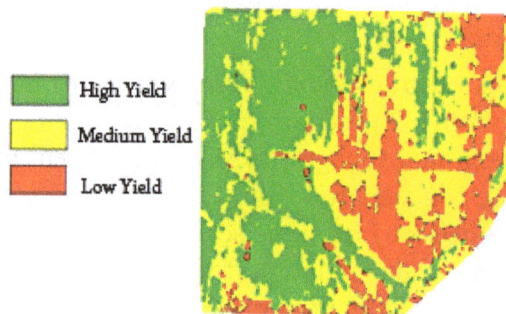

Agricultural Research Service scientists have developed a way to incorporate lidar with yield rates on agricultural fields. This technology will help farmers improve their yields by directing their resources toward the high-yield sections of their land.

Another application of lidar beyond crop health and terrain mapping is crop mapping in orchards and vineyards. Vehicles equipped with lidar sensors can detect foliage growth to determine if pruning or other maintenance needs to take place, detect variations in fruit production, or perform automated tree counts.

Lidar is useful in GPS-denied situations, such as in nut and fruit orchards where GPS signals to farm equipment featuring precision agriculture technology or a driverless tractor may be partially or completely blocked by overhanging foliage. Lidar sensors can detect the edges of rows so that farming equipment can continue moving until GPS signal can be reestablished.

Archaeology

Lidar has many applications in the field of archaeology including aiding in the planning of field campaigns, mapping features beneath forest canopy, and providing an overview of broad, continuous features that may be indistinguishable on the ground. Lidar can also provide archaeologists with the ability to create high-resolution digital elevation models (DEMs) of archaeological sites that can reveal micro-topography that are otherwise hidden by vegetation. Lidar-derived products can be easily integrated into a Geographic Information System (GIS) for analysis and interpretation. For example, at Fort Beauséjour - Fort Cumberland National Historic Site, Canada, previously undiscovered archaeological features below forest canopy have been mapped that are related to the siege of the Fort in 1755. Features that could not be distinguished on the ground or through aerial photography were identified by overlaying hillshades of the DEM created with artificial illumination from various angles. With lidar, the ability to produce high-resolution datasets quickly and relatively cheaply can be an advantage. Beyond efficiency, its ability to penetrate forest canopy has led to the discovery of features that were not distinguishable through traditional geo-spatial methods and are difficult to reach through field surveys, as in work at Caracol by Arlen Chase and his wife Diane Zaino Chase. The intensity of the returned signal can be used to detect features buried under flat vegetated surfaces such as fields, especially when mapping using the infrared spectrum. The presence of these features affects plant growth and thus the amount of infrared light reflected back. In 2012, lidar was used by a team attempting to find the legendary city of La Ciudad Blanca in the Honduran jungle. During a seven-day mapping period, they found evidence of extensive man-made structures that had eluded ground searches for hundreds of years. In June 2013 the rediscovery of the city of Mahendraparvata was announced. In another study, lidar was used to reveal stone walls, building foundations, abandoned roads, and other features of the landscape in southern New England, USA that had been obscured in aerial photography by the region's dense forest canopy. In May 2012, lidar was used to locate a previously unknown ruined city in the La Mosquitia region of Honduras.

Autonomous Vehicles

Forecast 3D Laser System uses a SICK LMC lidar sensor

Autonomous vehicles use lidar for obstacle detection and avoidance to navigate safely through environments, using rotating laser beams. Cost map or point cloud outputs from the lidar sensor provide the necessary data for robot software to determine where potential obstacles exist in the environment and where the robot is in relation to those potential obstacles. Singapore's *Singapore-MIT Alliance for Research and Technology (SMART)* is actively developing technologies for autonomous lidar vehicles. Examples of companies that produce lidar sensors commonly used in robotics or vehicle automation are Sick and Hokuyo. Examples of obstacle detection and avoidance products that leverage lidar sensors are the Autonomous Solution, Inc. Forecast 3D Laser System and Velodyne HDL-64E.

The very first generations of automotive adaptive cruise control systems used only lidar sensors.

It has been shown that lidar can be manipulated, such that self-driving cars are tricked into taking evasive action.

Biology and Conservation

Lidar has also found many applications in forestry. Canopy heights, biomass measurements, and leaf area can all be studied using airborne lidar systems. Similarly, lidar is also used by many industries, including Energy and Railroad, and the Department of Transportation as a faster way of surveying. Topographic maps can also be generated readily from lidar, including for recreational use such as in the production of orienteering maps.

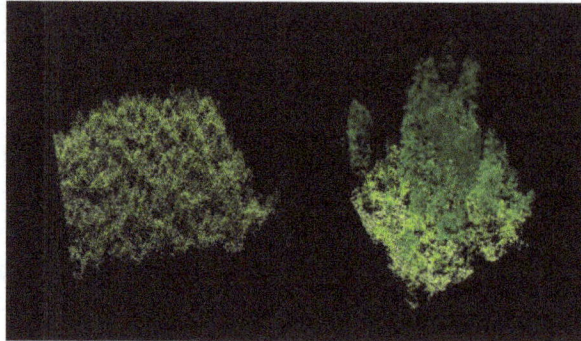

Lidar imaging comparing old-growth forest (right) to a new plantation of trees (left).

In addition, the Save-the-Redwoods League is undertaking a project to map the tall redwoods on the Northern California coast. Lidar allows research scientists to not only measure the height of previously unmapped trees but to determine the biodiversity of the redwood forest. Stephen Sillett, who is working with the League on the North Coast lidar project, claims this technology will be useful in directing future efforts to preserve and protect ancient redwood trees.

Geology and Soil Science

High-resolution digital elevation maps generated by airborne and stationary lidar have led to significant advances in geomorphology (the branch of geoscience concerned with the origin and evolution of the Earth surface topography). The lidar abilities to detect subtle topographic features such as river terraces and river channel banks, to measure the land-surface elevation beneath the vegetation canopy, to better resolve spatial derivatives of elevation, and to detect elevation changes between repeat surveys have enabled many novel studies of the physical and chemical processes that shape landscapes. In 2005 the Tour Ronde in the Mont Blanc massif became the first high alpine mountain on which lidar was employed to monitor the increasing occurrence of severe rockfall over large rock faces allegedly caused by climate change and degradation of permafrost at high altitude.

In geophysics and tectonics, a combination of aircraft-based lidar and GPS has evolved into an important tool for detecting faults and for measuring uplift. The output of the two technologies can produce extremely accurate elevation models for terrain - models that can even measure ground elevation through trees. This combination was used most famously to find the location of the Seattle Fault in Washington, United States. This combination also measures uplift at Mt. St. Helens by using data from before and after the 2004 uplift. Airborne lidar systems monitor glaciers and have the ability to detect subtle amounts of growth or decline. A satellite-based system, the NASA ICESat, includes a lidar sub-system for this purpose. The NASA Airborne Topographic Mapper is also used extensively to monitor glaciers and perform coastal change analysis. The combination is also used by soil scientists while creating a soil survey. The detailed terrain modeling allows soil scientists to see slope changes and landform breaks which indicate patterns in soil spatial relationships.

Atmospheric Remote Sensing and Meteorology

Initially based on ruby lasers, lidar for meteorological applications was constructed shortly after the invention of the laser and represent one of the first applications of laser technology. Lidar technology has since expanded vastly in capability and lidar systems are used to perform a range of measurements that include profiling clouds, measuring winds, studying aerosols and quantifying various atmospheric components. Atmospheric components can in turn provide useful information including surface pressure (by measuring the absorption of oxygen or nitrogen), greenhouse gas emissions (carbon dioxide and methane), photosynthesis (carbon dioxide), fires (carbon monoxide) and humidity (water vapor). Atmospheric lidars can be either ground-based, airborne or satellite depending on the type of measurement.

Atmospheric lidar remote sensing works in two ways -

1. by measuring backscatter from the atmosphere, and

2. by measuring the scattered reflection off the ground (when the lidar is airborne) or other hard surface.

Backscatter from the atmosphere directly gives a measure of clouds and aerosols. Other derived measurements from backscatter such as winds or cirrus ice crystals require careful selecting of the wavelength and/or polarization detected. *Doppler Lidar* and *Rayleigh Doppler Lidar* are used to measure temperature and/or wind speed along the beam by measuring the frequency of the backscattered light. The Doppler broadening of gases in motion allows the determination of properties via the resulting frequency shift. Scanning lidars, such as the conical-scanning NASA HARLIE LIDAR, have been used to measure atmospheric wind velocity. The ESA wind mission ADM-Aeolus will be equipped with a Doppler lidar system in order to provide global measurements of vertical wind profiles. A doppler lidar system was used in the 2008 Summer Olympics to measure wind fields during the yacht competition.

Doppler lidar systems are also now beginning to be successfully applied in the renewable energy sector to acquire wind speed, turbulence, wind veer and wind shear data. Both pulsed and continuous wave systems are being used. Pulsed systems use signal timing to obtain vertical distance resolution, whereas continuous wave systems rely on detector focusing.

The term *eolics* has been proposed to describe the collaborative and interdisciplinary study of wind using computational fluid mechanics simulations and Doppler lidar measurements.

The ground reflection of an airborne lidar gives a measure of surface reflectivity (assuming the atmospheric transmittance is well known) at the lidar wavelength. However, the ground reflection is typically used for making absorption measurements of the atmosphere. "Differential absorption lidar" (DIAL) measurements utilize two or more closely spaced (<1 nm) wavelengths to factor out surface reflectivity as well as other transmission losses, since these factors are relatively insensitive to wavelength. When tuned to the appropriate absorption lines of a particular gas, DIAL measurements can be used to determine the concentration (mixing ratio) of that particular gas in the atmosphere. This is referred to as an *Integrated Path Differential Absorption* (IPDA) approach, since it is a measure of the integrated absorption along the entire lidar path. IPDA lidars can be either pulsed or CW and typically use two or more wavelengths. IPDA lidars have been used for remote sensing of carbon dioxide and methane.

Synthetic array lidar allows imaging lidar without the need for an array detector. It can be used for imaging Doppler velocimetry, ultra-fast frame rate (MHz) imaging, as well as for speckle reduction in coherent lidar. An extensive lidar bibliography for atmospheric and hydrospheric applications is given by Grant.

Law Enforcement

Lidar speed guns are used by the police to measure the speed of vehicles for speed limit enforcement purposes.

Military

Few military applications are known to be in place and are classified (like the lidar-based speed measurement of the AGM-129 ACM stealth nuclear cruise missile), but a considerable amount of research is underway in their use for imaging. Higher resolution systems collect enough detail to identify targets, such as tanks. Examples of military applications of lidar include the Airborne Laser Mine Detection System (ALMDS) for counter-mine warfare by Areté Associates.

A NATO report (RTO-TR-SET-098) evaluated the potential technologies to do stand-off detection for the discrimination of biological warfare agents. The potential technologies evaluated were Long-Wave Infrared (LWIR), Differential Scattering (DISC), and Ultraviolet Laser Induced Fluorescence (UV-LIF). The report concluded that : *Based upon the results of the lidar systems tested and discussed above, the Task Group recommends that the best option for the near-term (2008–2010) application of stand-off detection systems is UV LIF* . However, in the long-term, other techniques such as stand-off Raman spectroscopy may prove to be useful for identification of biological warfare agents.

Short-range compact spectrometric lidar based on Laser-Induced Fluorescence (LIF) would address the presence of bio-threats in aerosol form over critical indoor, semi-enclosed and outdoor venues like stadiums, subways, and airports. This near real-time capability would enable rapid detection of a bioaerosol release and allow for timely implementation of measures to protect occupants and minimize the extent of contamination.

The Long-Range Biological Standoff Detection System (LR-BSDS) was developed for the US Army to provide the earliest possible standoff warning of a biological attack. It is an airborne system carried by a helicopter to detect man-made aerosol clouds containing biological and chemical agents at long range. The LR-BSDS, with a detection range of 30 km or more, was fielded in June 1997. Five lidar units produced by the German company Sick AG were used for short range detection on Stanley, the autonomous car that won the 2005 DARPA Grand Challenge.

A robotic Boeing AH-6 performed a fully autonomous flight in June 2010, including avoiding obstacles using lidar.

Mining

Lidar is used in the mining industry for various tasks. The calculation of ore volumes is accomplished by periodic (monthly) scanning in areas of ore removal, then comparing surface data to the previous scan.

Lidar sensors may also be used for obstacle detection and avoidance for robotic mining vehicles such as in the Komatsu Autonomous Haulage System (AHS) used in Rio Tinto's Mine of the Future.

Physics and Astronomy

A worldwide network of observatories uses lidars to measure the distance to reflectors placed on the moon, allowing the position of the moon to be measured with mm precision and tests of general relativity to be done. MOLA, the Mars Orbiting Laser Altimeter, used a lidar instrument in a Mars-orbiting satellite (the NASA Mars Global Surveyor) to produce a spectacularly precise global topographic survey of the red planet.

In September, 2008, the NASA Phoenix Lander used lidar to detect snow in the atmosphere of Mars.

In atmospheric physics, lidar is used as a remote detection instrument to measure densities of certain constituents of the middle and upper atmosphere, such as potassium, sodium, or molecular nitrogen and oxygen. These measurements can be used to calculate temperatures. Lidar can also be used to measure wind speed and to provide information about vertical distribution of the aerosol particles.

At the JET nuclear fusion research facility, in the UK near Abingdon, Oxfordshire, lidar Thomson Scattering is used to determine Electron Density and Temperature profiles of the plasma.

Rock Mechanics

LiDAR has been widely used in rock mechanics for rock mass characterization and slope change detection. Some important geomechanical properties from the rock mass can be extracted from the 3D point clouds obtained by means of the LiDAR. Some of these properties are:

- Discontinuity orientation
- Discontinuity spacing and RQD
- Discontinuity aperture
- Discontinuity persistence
- Discontinuity roughness
- Water infiltration

Some of these properties have be used to assess the geomechanical quality of the rock mass through the RMR index. Moreover, as the orientations of discontinuities can be extracted using the existing methodologies, it is possible to assess the geomechanical quality of a rock slope through the SMR index. In addition to this, the comparison of different 3D point clouds from a slope acquired at different times allows to study the changes produced on the scene during this time interval as a result of rockfalls or any other landsliding processes.

Robotics

Lidar technology is being used in robotics for the perception of the environment as well as object classification. The ability of lidar technology to provide three-dimensional elevation maps of the terrain, high precision distance to the ground, and approach velocity can enable safe landing of robotic and manned vehicles with a high degree of precision. Refer to the Military section above for further examples.

Spaceflight

Lidar is increasingly being utilized for rangefinding and orbital element calculation of relative velocity in proximity operations and stationkeeping of spacecraft. Lidar has also been used for atmospheric studies from space. Short pulses of laser light beamed from a spacecraft can reflect off of tiny particles in the atmosphere and back to a telescope aligned with the spacecraft laser. By precisely timing the lidar 'echo,' and by measuring how much laser light is received by the telescope, scientists can accurately determine the location, distribution and nature of the particles. The result is a revolutionary new tool for studying constituents in the atmosphere, from cloud droplets to industrial pollutants, that are difficult to detect by other means."

Surveying

Airborne lidar sensors are used by companies in the remote sensing field. They can be used to create a DTM (Digital Terrain Model) or DEM (Digital Elevation Model); this is quite a common practice for larger areas as a plane can acquire 3–4 km wide swaths in a single flyover. Greater vertical accuracy of below 50 mm can be achieved with a lower flyover, even in forests, where it is able to give the height of the canopy as well as the ground elevation. Typically, a GNSS receiver configured over a georeferenced control point is needed to link the data in with the WGS (World Geodetic System).

This TomTom mapping van is fitted with five lidars on its roof rack.

Transport

LiDAR has been used in the railroad industry to generate asset health reports for asset management and by departments of transportation to assess their road conditions. CivilMaps.com is a leading company in the field. Lidar has been used in adaptive cruise control (ACC) systems for automobiles. Systems such as those by Siemens and Hella use a lidar device mounted on the front of the vehicle, such as the bumper, to monitor the distance between the vehicle and any vehicle in front of it. In the event the vehicle in front slows down or is too close, the ACC applies the brakes to slow the vehicle. When the road ahead is clear, the ACC allows the vehicle to accelerate to a speed preset by the driver. Refer to the Military section above for further examples.

Wind Farm Optimization

Lidar can be used to increase the energy output from wind farms by accurately measuring wind speeds and wind turbulence. Experimental lidar systems can be mounted on the nacelle of a wind turbine or integrated into the rotating spinner to measure oncoming horizontal winds, winds in the wake of the wind turbine, and proactively adjust blades to protect components and increase power. Lidar is also used to characterise the incident wind resource for comparison with wind turbine power production to verify the performance of the wind turbine by measuring the wind turbine's power curve. Wind farm optimization can be considered a topic in *applied eolics*.

Solar Photovoltaic Deployment Optimization

Lidar can also be used to assist planners and developers in optimizing solar photovoltaic systems at the city level by determining appropriate roof tops and for determining shading losses. Recent works focus on buildings' facades solar potential estimation, or by incorporating more detailed shading losses by considering the influence from vegetation and larger surrounding terrain.

Video Games

Racing game iRacing features scanned tacks, resulting in bumps with millimeter precision in the in-game 3D mapping environment.

Other Uses

The video for the song "House of Cards" by Radiohead was believed to be the first use of real-time 3D laser scanning to record a music video. The range data in the video is not completely from a lidar, as structured light scanning is also used.

Alternative Technologies

Recent development of Structure From Motion (SFM) technologies allows delivering 3D images and maps based on data extracted from visual and IR photography. The elevation or 3D data is extracted using multiple parallel passes over mapped area, yielding both visual light image and 3D structure from the same sensor, which is often a specially chosen and calibrated digital camera.

ERDAS Imagine

ERDAS Imagine is a remote sensing application with raster graphics editor abilities designed by ERDAS for geospatial applications. The latest version is 2015. Imagine is aimed mainly at geospatial raster data processing and allows users to prepare, display and enhance digital images for mapping use in geographic information system (GIS) and computer-aided design (CAD) software. It is a toolbox allowing the user to perform numerous operations on an image and generate an answer to specific geographical questions.

By manipulating imagery data values and positions, it is possible to see features that would not normally be visible and to locate geo-positions of features that would otherwise be graphical. The level of brightness, or reflectance of light from the surfaces in the image can be helpful with vegetation analysis, prospecting for minerals etc. Other usage examples include linear feature extraction, generation of processing work flows (*spatial models* in Imagine), import/export of data for a wide variety of formats, orthorectification, mosaicking of imagery, stereo and automatic feature extraction of map data from imagery.

Product History

Before the ERDAS Imagine Suite, ERDAS, Inc. developed various products to process satellite imagery from AVHRR, Landsat MSS and TM, and Spot Image into land cover, land use maps, map deforestation, and assist in locating oil reserves under the product name ERDAS. These older ERDAS applications were rewritten from Fortran to C and C++ and exist today within the Imagine Suite, which has grown to support most optical and radar mapping satellites, airborne mapping cameras and digital sensors used for mapping.

ERDAS 4

The first version of ERDAS was launched in 1978 on Cromemco microcomputers based on the 8-bit Z80 CPU running the CDOS operating system. The system was built into a desk and was configured with one color monitor (256 x 256 resolution), one B&W monitor, two 8" floppy drives (one for software and one for data). Later, other options were added, such as a large digitizing tablet, and a hard drive (hard drives usually did not exist on computers this size). The hard drive was a Control Data Corporation model about the size of a small washing machine. It had 80 MB fixed disks and a very large 16 MB removable platter. When the hard disk was installed, ERDAS engineers had to cut out the back of the computer furniture to make room for the cables.

ERDAS 400

In 1980, ERDAS, Inc. developed the ERDAS 400. This early system was a turnkey computerized forest-management system for the state of New York. The hardware needed to make this ability available at an affordable price did not exist at the time, therefore some circuit board and other hardware designs were created by ERDAS to deliver the product. Between 1980 and 1982 different versions of ERDAS 400 legacy were delivered to NASA, US Forest Service, US Environmental Protection Agency, the State of Illinois, and others.

ERDAS 7.x

In November 1982 ERDAS 7.0 was released. This product was a change from the more turnkey, custom design of the ERDAS 400 systems to a DOS IBM Personal Computer. These systems continued to be command prompt based, but added a nested menu assisting customers through prompts needed to process the imagery into data used in a GIS.

In ERDAS 7.2 (Jan 1983) the first PC version of ERDAS software was licensed to the University of South Carolina Department of Geography. The release of versions ERDAS 7.0 - 7.5 marked a period of wide acceptance of remote sensing technology among US federal and state government agencies, and research universities needing to monitor natural resource changes.

A link between ESRI ARC/INFO and ERDAS 7.3 was introduced in 1988. The "ERDAS-ARC/INFO Live Link" allowed the mapping community to use technology from both companies to deliver high quality image display and image processing with powerful GIS ability. This relationship between ESRI and ERDAS, Inc. remained strong at a high management level and in various product offerings into 2007. This strong relationship began very early, as the ERDAS founders Lawrie Jordan and Bruce Rado and ESRI founder Jack Dangermond all attended the Harvard Graduate School of Design.

The final version of the ERDAS product line was ERDAS 7.5, which was finished as the new product, ERDAS Imagine, was being developed. All the abilities of ERDAS 7.5 were merged into Imagine.

ERDAS Imagine

ERDAS Imagine was demonstrated in October 1991 and released as ERDAS Imagine 8.0 in February 1992. It was released on a Sun Workstation using SunOS providing a graphical user interface (GUI) to assist in visualizing imagery used in mapping, vector GIS data, creating maps, and so forth. Much of the ERDAS 7.5 product was provided for several years through intermediate product releases of ERDAS 8.01 & 8.02 until the ERDAS Imagine product replaced all the ERDAS 7.5 ability in 1994 with the release of Imagine 8.1.

The relationship between the ERDAS and ARC/INFO products began with the "ERDAS-ARC/Info Live Link" and was expanded in ERDAS Imagine 8.2 (Feb 1993), when ERDAS released the Imagine Vector Module. This add-on was fully developed by ERDAS, but licensed the Arc Coverage data format from ESRI. The vector module is one of the most popular modules within the Imagine product suite through the 9.3 release (Sept 2008).

The processing of radar data in ERDAS Imagine began with the development of the Radar Module, first released in Nov. 1992. The abilities have been upgraded many times since with the original tools now named *Radar Interpreter*.

ERDAS Imagine Spatial Modeler

With the release of ERDAS 7.5 in 1990 ERDAS introduced a GIS Modeling (GISMO) script language. Tools were made available to assist in the development of complex spatial models using Dana Tomlin's map algebra concepts. GISMO was upgraded in 1992 with the release of ERDAS Imagine 8.0 to the Spatial Modeler scripting language. Model Maker, a graphic flow chart model building enhancement to Spatial Modeler was introduced in 1993. In 2004 ESRI copied the graphic flow chart model building environment idea and released an ArcGIS component named Model Builder.

With the release of ERDAS Imagine 2013 in December 2012, the Model Maker's graphical user interface was changed and its architecture was rewritten from a push model to a pull model. This re-architecture allows many complex models to be processed directly to a viewer in near-real time. A Python capability was added at this time.

A variety of ready to go spatial models can be downloaded from Sterling Geo's website who distribute the software in the UK.

TerrSet

TerrSet (formerly IDRISI) is an integrated geographic information system (GIS) and remote sensing software developed by Clark Labs at Clark University for the analysis and display of digital geospatial information. TerrSet is a PC grid-based system that offers tools for researchers and scientists engaged in analyzing earth system dynamics for effective and responsible decision making for environmental management, sustainable resource development and equitable resource allocation.

Key features of TerrSet include:

- a complete GIS analysis package for basic and advanced spatial analysis, including tools for surface and statistical analysis, decision support, land change and prediction, and image time series analysis;

- a complete Image Processing system with extensive hard and soft classifers, including machine learning classifiers such as neural networks and classification tree analysis, as well as image segmentation for classification;

- Land Change Modeler, a land planning and decision support toolset that addresses the complexities of land change analysis and land change prediction.

- Habitat and Biodiversity Modeler, a modeling environment for habitat assessment and biodiversity modeling.

- Ecosystem Services Modeler, a spatial decision support system for assessing the value of natural capital.

- Earth Trends Modeler, an integrated suite of tools for the analysis of image time series to assess climate trends and impacts.

- Climate Change Adaptation Modeler, a facility for modeling future climate and its impacts.

- GeOSIRIS-REDD, a national-level REDD planning tool to assess deforestation, carbon emissions, agricultural revenue and carbon payments.

History and Background

IDRISI was first conceived in 1987 by Prof. J. Ronald Eastman of Clark University, Department of Geography, as an accessible yet robust PC-based GIS. Dr. Eastman continues to be the prime developer and chief architect of the software. In January 2015 Clark Labs released the TerrSet Geospatial Monitoring and Modeling software, version 18. Besides its primary research and scientific focus, TerrSet is popular as an academic tool for teaching the principal theories behind GIS at colleges and universities.

Since 1987, IDRISI, and now TerrSet, has been used by professionals in a wide range of industries in more than 180 countries worldwide. Environmental managers and researchers benefit from the unsurpassed range of geospatial tools—over 300 modules for the analysis and display of digital spatial information.

Based within the Graduate School of Geography at Clark University, Clark Labs and its software tools are known for pioneering advancements in areas such as decision support, uncertainty management, classifier development, change and time series analysis, and dynamic modeling. Partnering with such organizations as The Gordon and Betty Moore Foundation, Google.org, USDA, the United Nations, Conservation International, Imazon and Wildlife Conservation Society.

The software is named after cartographer Muhammad al-Idrisi (1100–1166).

In 2015, Clark Labs released a new product called TerrSet. TerrSet greatly expands the modeling capacity of the base IDRISI toolset.

Remote Sensing (Archaeology)

Remote sensing techniques in archaeology are an increasingly important component of the technical and methodological tool set available in archaeological research. The use of remote sensing techniques allows archaeologists to uncover unique data that is unobtainable using traditional archaeological excavation techniques.

Ground-based geophysical methods such as Ground Penetrating Radar and Magnetometry are also used for archaeological imaging. Although these are sometimes classed as remote sensing, they are usually considered a separate discipline.

Satellite Archaeology

Satellite archaeology is an emerging field of archaeology that uses high resolution satellites with thermal and infrared capabilities to pinpoint potential sites of interest in the earth around a meter or so in depth. The infrared light used by these satellites have longer wavelengths than that of visible light and are therefore capable of penetrating the earth's surface. The images are then taken and processed by an archaeologist who specializes in satellite remote sensing in order to find any subtle anomalies on the earth's surface.

Landscape features such as soil, vegetation, geology, and man-made structures of possible cultural interest have specific signatures that the multi-spectral satellites can help to identify. The satellites can then make a 3D image of the area to show if there are any man-made structures beneath soil and vegetation that can not be seen by the naked eye. Commercially available satellites have a .4m-90m resolution that make it possible to see most ancient sites and their associated features in such places as Egypt, Perù and Mexico. It is a hope of archaeologists that in the next few decades resolutions will improve to the point where they are capable of zooming in on a single pottery shard buried beneath the earth's surface.

Satellite archaeology is a non-invasive method for mapping and monitoring potential archaeological sites in an ever changing world that faces issues such as urbanization, looting, and groundwater pollution that could put pose threats to such sites. In spite of this, satellites in archaeology are mostly a tool for broad scale survey and focused excavation. All archaeological projects need ground work in order to verify any potential findings.

Examples of Regional Applications

Maya Research

Some of the most prominent remote sensing research has been done in regard to Maya studies in Mesoamerica. The Petén region of northern Guatemala is of particular focus because remote sensing technology is of very definite use there. The Petén is a densely forested region and it lacks modern settlements and infrastructure. As a result, it is extremely difficult to survey, and because of this remote sensing offers a solution to this research problem. The use of remote sensing techniques in this region is a great example of the applications these methods have for archaeologists. The Petén is a hilly, karstic, thickly forested landscape which offers an incredible barrier for field archaeologists to penetrate. With the advent of remote sensing techniques, a plethora of information has been uncovered about the region and about the people that inhabited it.

The Petén is arguably one of the most difficult of the Maya landscapes in which to subsist. It is questions regarding subsistence patterns and related problems that have driven remote sensing methodology in the hopes of understanding the complex adaptations that the Maya developed. Remote sensing methods have also proven invaluable when working to discover features, cisterns, and temples. Archaeologists have identified vegetative differentiation associated with such features. With the advent of remote sensing, archaeologists are able to pinpoint and study the features hidden beneath this canopy without ever visiting the jungle.

A pioneer in the use of remote sensing in Maya research is NASA archaeologist Tom Sever, who has applied remote sensing to research in Maya site discovery as well as mapping causeways (*sacbeob*) and roads. Sever has stressed the enormous use of remote sensing in uncovering settlement patterns, population densities, societal structure, communication, and transportation. Sever has done much of his research in the Petén region of northern Guatemala, where he and his research team have used satellite imagery and GIS to map undiscovered roads and causeways the ancient Maya built to connect cities and settlements. These landscape artifacts represent the advantage of using remote sensing as these causeways are not visible from the ground. By mapping these forms, Sever is able to locate new sites and further uncover ancient Maya methods of communicated and transportation. Sever and his team also use remote sensing methods to gather data on deforestation. The rain forests of the Petén are undergoing massive deforestation, and Sever's remote sensing offers another window into this understanding and halting this problem. Monitoring the rate of deforestation not only has important ecological value, but the use of remote sensing can detect landscape change. By measuring the magnitude of landscape change in terms of vegetative cover and soil geography, as well as shifting land use patterns and the associated cultural diversity, archaeologists are given a window into depletion rates and trends in anthropogenic landscape alteration.

Much attention has been devoted to the mapping of canals and irrigation systems. Synthetic Aperture Radar (SAR) has proved particularly useful in this research. SAR is a type of radar that is sensitive to linear and geometric features on the ground. It is also important to include a method called ground truthing, or the process of physically visiting (on foot) the localities surveyed to verify the data and help inform the interpretation. GPS is often used to aid in this process.

Ground-based geophysical methods have also been employed in Maya research. Ground Penetrating Radar (GPR) has been performed on a number of sites, including Chichen Itza. The GPR research has detected buried causeways and structures that might have otherwise gone unnoticed.

Maya "Collapse"

One of Sever's research goals is understanding the comparatively sudden decline of many Maya centers in the central Lowlands region by the end of the 1st millennium CE, a happenstance often referred to as the "(Classic) Maya collapse". Sever's research on communication and transportation systems points to an extensive societal infrastructure capable of supporting the building and maintenance of the causeways and roadways. Using satellite imagery, researchers have been able to map canals and reservoirs. These offer a glimpse into Maya cultural adaptations during the period of their highest population density. At the height of the classic period, the population in the Maya lowlands was 500 - 1300 people per square mile in rural areas, and even more in urban regions. This far outweighs the carrying capacity for this region, but this follows centuries of successful adaptation. Other data shows that by the end of the classic period, the Maya had already depleted much of the rain forest. Understanding how the ancient Maya adapted to this karst topography could shed light on solutions to modern ecological problems that modern peoples in the Petén currently face, which is much the same, except there are fewer people who are causing even more damage to the biodiversity and cultural diversity. Sever believes that the Maya collapse was a primarily ecological disaster. By detecting deforestation rates and trends can help us to understand how these same processes affected the Maya. An important contribution to the study of Maya has been provided by LiDAR thanks to its ability to penetrate dense tropical canopies. LiDAR has been applied to the site of Caracol, Belize in 2009, revaling an impressive monumental complex covered by jungle.

Satellite Archaeology in Peru

In Peru, an Italian scientific mission of CNR, directed by Nicola Masini, provided important results by using satellite imagery for both site discovery and the protection of archaeological heritage. In particular, by processing QuickBird images a large buried settlement, including a pyramid, in the Nasca riverbed (Southern Peru), near the Ceremonial Center of Cahuachi, has been detected. In the region of Lambayeque (Northern Peru), which is strongly affected by clandestine excavations, satellite imagery have been also employed for mapping and monitoring archaeological looting.

Location of Ancient Iram

Iram of the Pillars is a lost city (or region surrounding the lost city) on the Arabian Peninsula. In the early 1980s a group of researchers interested in the history of Iram used NASA remote sensing satellites, ground penetrating radar, Landsat program data and images taken from the Space Shuttle Challenger as well as SPOT data to identify old camel train routes and points where they converged. These roads were used as frankincense trade routes around 2800 BC to 100 BC.

One area in the Dhofar province of Oman was identified as a possible location for an outpost of the lost civilization. A team including adventurer Ranulph Fiennes, archaeologist Juris Zarins, filmmaker Nicholas Clapp, and lawyer George Hedges, scouted the area on several trips, and stopped at a water well called Ash Shisar. Near this oasis was located a site previously identified as the 16th century Shis'r fort. Excavations uncovered an older settlement, and artifacts traded from far and wide were found. This older fort was found to have been built on top of a large limestone cavern which would have served as the water source for the fort, making it an important oasis on the trade route to Iram. As the residents of the fort consumed the water from underground, the water table fell, leaving the limestone roof and walls of the cavern dry. Without the support of the water, the cavern would have been in danger of collapse, and it seems to have done so some time between 300-500 AD, destroying the oasis and covering over the water source.

Four subsequent excavations were conducted by Dr. Juris Zarins, tracing the historical presence by the people of 'Ad, the assumed ancestral builders of Iram.

Egypt and The Roman Empire

Archaeologist Dr Sarah Parcak uses satellites to search for sub-surface remains. Parcak uses these satellites to hunt to for lost settlements, tombs, and pyramids in Egypt's Nile Delta. She has also prospectively identified several significant sites in various parts of the ancient Roman Empire.

References

- Schowengerdt, Robert A. (2007). Remote sensing: models and methods for image processing (3rd ed.). Academic Press. p. 2. ISBN 978-0-12-369407-2.

- Liu, Jian Guo & Mason, Philippa J. (2009). Essential Image Processing for GIS and Remote Sensing. Wiley-Blackwell. p. 4. ISBN 978-0-470-51032-2.

- Cracknell, Arthur P.; Hayes, Ladson (2007) [1991]. Introduction to Remote Sensing (2 ed.). London: Taylor and Francis. ISBN 0-8493-9255-1. OCLC 70765252.

- "Lecture 14 : Principles of active remote sensing: Lidars and lidar sensing of aerosols, gases and clouds." (PDF). Laser-distance-measurer.com. Retrieved 2016-02-22.

- By Steve Taranovich, EDN. "Autonomous automotive sensors: How processor algorithms get their inputs." July 5, 2016. Retrieved August 9, 2016.

- Douglas Preston (2 Mar 2015). "Lost City Discovered in the Honduran Rain Forest". National Geographic. Retrieved 3 March 2015.

- "Filipino turns ordinary car into autonomous vehicle - Motioncars | Motioncars". Motioncars.inquirer.net. 2015-05-25. Retrieved 2016-02-22.

- Gallacher, D., and More, G., Lidar measurements and visualisation of turbulence and wake decay length European Wind Energy Association Annual Conference, 2014. Retrieved: 9 May 2014.

- Clive, P. J. M., Offshore power performance assessment for onshore costs DEWEK (Deutsche Windenergie Konferenz), 2012. Retrieved: 9 May 2014.

- Talbot, David (2014-02-13). "New Optical Chip Will Sharpen Military and Archeological Aerial Imaging | MIT

Technology Review". Technologyreview.com. Retrieved 2014-02-17.

- "The world's first control of a wind turbine with a nacelle-based Lidar system". Corporate Communications University of Stuttgart. 2012-06-05. Retrieved 2014-04-12.

- "LIDAR—Light Detection and Ranging—is a remote sensing method used to examine the surface of the Earth". NOAA. Archived from the original on June 4, 2013. Retrieved June 4, 2013.

- First Name Middle Name Last Name (2011-09-27). "IEEE Xplore - An overview of LITE: NASA's Lidar In-space Technology Experiment" (PDF). Ieeexplore.ieee.org. doi:10.1109/5.482227. Retrieved 2013-05-06.

- "Dr. J. Ronald Eastman to be awarded the Distinguished Career Award at 2010 Annual AAG Meeting". Directions Magazine. Directions Media. Retrieved 28 September 2012.

- "Clark Labs Receives $1.8 Million Grant from the Moore Foundation to Develop Land Management Software". Directions Magazine. Directions Media. Retrieved 28 September 2012.

- "Clark Labs Receives Support from Google.org to Develop On-Line Prototype of its Land Change Modeler Application to be run on Google's Earth Engine". Directions Magazine. Directions Media. Retrieved 28 September 2012.

- "Clark Labs and Conservation International Partner to Develop REDD-Specific Tools within IDRISI Taiga's Land Change Modeler Application". Directions Magazine. Directions Media. Retrieved 28 September 2012.

- "Clark Labs at Clark University Strengthens Collaboration with Wildlife Conservation Society through Teaching and Research Partnership". Directions Magazine. Directions Media. Retrieved 28 September 2012.

- John Nobel Wilford (2010-05-10). "Mapping Ancient Civilization, in a Matter of Days". New York Times. Retrieved 2010-05-11.

- Mikkelsen, Torben & Hansen, Kasper Hjorth et al. Lidar wind speed measurements from a rotating spinner Danish Research Database & Danish Technical University, 20 April 2010. Retrieved: 25 April 2010.

Radar and its Applications

Countries use radar to detect aircrafts, missiles, vehicles and spacecrafts. It is a system that uses radio waves to determine the objects in the area concerned. It has proven to be of great importance to the military complex. The diverse applications of radars in the current scenario have been thoroughly discussed in this chapter.

Radar

Radar is an object-detection system that uses radio waves to determine the range, angle, or velocity of objects. It can be used to detect aircraft, ships, spacecraft, guided missiles, motor vehicles, weather formations, and terrain. A radar system consists of a transmitter producing electromagnetic waves in the radio or microwaves domain, an emitting antenna, a receiving antenna (separate or the same as the previous one) to capture any returns from objects in the path of the emitted signal, a receiver and processor to determine properties of the object(s).

Long-range radar antenna, used to track space objects and ballistic missiles.

Radar was secretly developed by several nations in the period before and during World War II. The term *RADAR* was coined in 1940 by the United States Navy as an acronym for RAdio Detection And Ranging. The term *radar* has since entered English and other languages as a common noun, losing all capitalization.

The modern uses of radar are highly diverse, including air and terrestrial traffic control, radar astronomy, air-defense systems, antimissile systems, marine radars to locate landmarks and other ships, aircraft anticollision systems, ocean surveillance systems, outer space surveillance and rendezvous systems, meteorological precipitation monitoring, altimetry and flight control systems, guided missile target locating systems, ground-penetrating radar for geological observations, and range-controlled radar for public health surveillance. High tech radar systems are associated with digital signal processing, machine learning and are capable of extracting useful information from very high noise levels.

Radar of the type used for detection of aircraft. It rotates steadily, sweeping the airspace with a narrow beam.

Other systems similar to radar make use of other parts of the electromagnetic spectrum. One example is "lidar", which uses ultraviolet, visible, or near infrared light from lasers rather than radio waves.

History

As early as 1886, German physicist Heinrich Hertz showed that radio waves could be reflected from solid objects. In 1895, Alexander Popov, a physics instructor at the Imperial Russian Navy school in Kronstadt, developed an apparatus using a coherer tube for detecting distant lightning strikes. The next year, he added a spark-gap transmitter. In 1897, while testing this equipment for communicating between two ships in the Baltic Sea, he took note of an interference beat caused by the passage of a third vessel. In his report, Popov wrote that this phenomenon might be used for detecting objects, but he did nothing more with this observation.

The German inventor Christian Hülsmeyer was the first to use radio waves to detect "the presence of distant metallic objects". In 1904 he demonstrated the feasibility of detecting a ship in dense fog, but not its distance from the transmitter. He obtained a patent for his detection device in April 1904 and later a patent for a related amendment for estimating the distance to the ship. He also got a British patent on September 23, 1904 for a full radar system, that he called a *telemobiloscope*. It operated on a 50 cm wavelength and the pulsed radar signal was created via a spark-gap. His system already used the classic antenna setup of horn antenna with parabolic reflector and was presented to German military officials in practical tests in Cologne and Rotterdam harbour but was rejected.

In 1922 A. Hoyt Taylor and Leo C. Young, researchers working with the U.S. Navy, had a transmitter and a receiver on opposite sides of the Potomac River and discovered that a ship passing through the beam path caused the received signal to fade in and out. Taylor submitted a report, suggesting that this might be used to detect the presence of ships in low visibility, but the Navy did not immediately continue the work. Eight years later, Lawrence A. Hyland at the Naval Research Laboratory observed similar fading effects from a passing aircraft; this led to a patent application as well as a proposal for serious work at the NRL (Taylor and Young were then at this laboratory) on radio-echo signals from moving targets.

Experimental radar antenna, US Naval Research Laboratory, Anacostia, D. C., late 1930s

During the 1920s the UK research establishment made many advances using radio techniques, including the probing of the ionosphere and the detection of lightning at long distances. Robert Watson-Watt became an expert on the use of radio direction finding as part of his lightning experiments. As part of ongoing experiments, he asked the "new boy", Arnold Frederic Wilkins, to find a receiver suitable for use with shortwave transmissions. Wilkins made an extensive study of available units before selecting a model from the General Post Office. Its instruction manual noted that there was "fading" (the common term for interference at the time) when aircraft flew by.

A Chain Home tower in Great Baddow, United Kingdom

Before the Second World War, researchers in France, Germany, Italy, Japan, the Netherlands, the Soviet Union, the United Kingdom, and the United States, independently and in great secrecy, developed technologies that led to the modern version of radar. Australia, Canada, New Zealand, and South Africa followed prewar Great Britain, and Hungary had similar developments during the war.

In France in 1934, following systematic studies on the Split Anode Magnetron, the research branch of the Compagnie Générale de Télégraphie Sans Fil (CSF), headed by Maurice Ponte, with Henri Gutton, Sylvain Berline, and M. Hugon, began developing an obstacle-locating radio apparatus, a part of which was installed on the liner *Normandie* in 1935.

During the same time, the Soviet military engineer P. K. Oshchepkov, in collaboration with Leningrad Electrophysical Institute, produced an experimental apparatus, RAPID, capable of detecting an aircraft within 3 km of a receiver. The Soviets produced their first mass production radars RUS-1 and RUS-2 Redut in 1939 but further development was slowed by the NKVD arrest of Oshchepkov and their sending him to the gulag. In total, only 607 Redut stations were produced during the war. The first Russian airborne radar, Gneiss-2, entered into service in June 1943 on Pe-2 fighters. More than 230 Gneiss-2 stations were produced by the end of 1944. The French and Soviet systems, however, had continuous-wave operation and could not give the full performance that was ultimately at the center of modern radar.

Full radar evolved as a pulsed system, and the first such elementary apparatus was demonstrated in December 1934 by the American Robert M. Page, working at the Naval Research Laboratory. The following year, the United States Army successfully tested a primitive surface-to-surface radar to aim coastal battery search lights at night. This was followed by a pulsed system demonstrated in May 1935 by Rudolf Kühnhold and the firm GEMA in Germany and then one in June 1935 by an Air Ministry team led by Robert A. Watson-Watt in Great Britain. Development of radar greatly expanded on 1 September 1936 when Watson-Watt became Superintendent of a new establishment under the British Air Ministry, Bawdsey Research Station located in Bawdsey Manor, near Felixstowe, Suffolk. Work there resulted in the design and installation of aircraft detection and tracking stations called "Chain Home" along the East and South coasts of England in time for the outbreak of World War II in 1939. This system provided the vital advance information that helped the Royal Air Force win the Battle of Britain.

In 1935 Watt was asked to pass judgement on recent reports of a German radio-based death ray and turned the request over to Wilkins. Wilkins returned a set of calculations demonstrating the system was basically impossible. When Watt then asked what might they do, Wilkins recalled the earlier report about aircraft causing radio interference. This led to the Daventry Experiment of February 26, 1935, using a powerful BBC shortwave transmitter as the source and their GPO receiver set up in a field while a bomber flew around the site. When returns were clearly seen, funds were immediately provided for development of an operational system. Watt's team patented the device in GB593017.

Given all required funding and development support, the team had working radar systems in 1935 and began deployment. By 1936 the first five Chain Home (CH) systems were operational and by 1940 stretched across the entire UK including Northern Ireland. Even by standards of the era, CH was crude; instead of broadcasting and receiving from an aimed antenna, CH broadcast a signal floodlighting the entire area in front of it, and then used one of Watt's own radio direction finders to determine the direction of the returned echoes. This meant that CH transmitters had to be much more powerful and have better antennas than competing systems but allowed its rapid introduction using existing technologies.

In April 1940, *Popular Science* showed an example of a radar unit using the Watson-Watt patent in an article on air defence. Also, in late 1941 *Popular Mechanics* had an article in which a U.S. scientist speculated about the British early warning system on the English east coast and came close to what it was and how it worked. Alfred Lee Loomis organized the Radiation Laboratory at Cambridge, Massachusetts which developed the technology in the years 1941-45. Later, in 1943, Page greatly improved radar with the monopulse technique that was used for many years in most radar applications.

The war precipitated research to find better resolution, more portability, and more features for radar, including complementary navigation systems like Oboe used by the RAF's Pathfinder.

Applications

The information provided by radar includes the bearing and range (and therefore position) of the object from the radar scanner. It is thus used in many different fields where the need for such positioning is crucial. The first use of radar was for military purposes: to locate air, ground and sea targets. This evolved in the civilian field into applications for aircraft, ships, and roads.

Commercial marine radar antenna. The rotating antenna radiates a vertical fan-shaped beam.

In aviation, aircraft are equipped with radar devices that warn of aircraft or other obstacles in or approaching their path, display weather information, and give accurate altitude readings. The first commercial device fitted to aircraft was a 1938 Bell Lab unit on some United Air Lines aircraft. Such aircraft can land in fog at airports equipped with radar-assisted ground-controlled approach systems in which the plane's flight is observed on radar screens while operators radio landing directions to the pilot.

Marine radars are used to measure the bearing and distance of ships to prevent collision with other ships, to navigate, and to fix their position at sea when within range of shore or other fixed references such as islands, buoys, and lightships. In port or in harbour, vessel traffic service radar systems are used to monitor and regulate ship movements in busy waters.

Meteorologists use radar to monitor precipitation and wind. It has become the primary tool for short-term weather forecasting and watching for severe weather such as thunderstorms, tornadoes, winter storms, precipitation types, etc. Geologists use specialized ground-penetrating radars to map the composition of Earth's crust.

Police forces use radar guns to monitor vehicle speeds on the roads.

Principles

Radar Signal

A radar system has a transmitter that emits radio waves called *radar signals* in predetermined directions. When these come into contact with an object they are usually reflected or scattered in many directions. Radar signals are reflected especially well by materials of considerable electrical conductivity—especially by most metals, by seawater and by wet ground. Some of these make the use of radar altimeters possible. The radar signals that are reflected back towards the transmitter are the desirable ones that make radar work. If the object is *moving* either toward or away from the transmitter, there is a slight equivalent change in the frequency of the radio waves, caused by the Doppler effect.

Radar receivers are usually, but not always, in the same location as the transmitter. Although the reflected radar signals captured by the receiving antenna are usually very weak, they can be strengthened by electronic amplifiers. More sophisticated methods of signal processing are also used in order to recover useful radar signals.

The weak absorption of radio waves by the medium through which it passes is what enables radar sets to detect objects at relatively long ranges—ranges at which other electromagnetic wavelengths, such as visible light, infrared light, and ultraviolet light, are too strongly attenuated. Such weather phenomena as fog, clouds, rain, falling snow, and sleet that block visible light are usually transparent to radio waves. Certain radio frequencies that are absorbed or scattered by water vapor, raindrops, or atmospheric gases (especially oxygen) are avoided in designing radars, except when their detection is intended.

Illumination

Radar relies on its own transmissions rather than light from the Sun or the Moon, or from electromagnetic waves emitted by the objects themselves, such as infrared wavelengths (heat). This process of directing artificial radio waves towards objects is called *illumination*, although radio waves are invisible to the human eye or optical cameras.

Reflection

If electromagnetic waves traveling through one material meet another material, having a different dielectric constant or diamagnetic constant from the first, the waves will reflect or scatter from the boundary between the materials. This means that a solid object in air or in a vacuum, or a significant change in atomic density between the object and what is surrounding it, will usually scatter radar (radio) waves from its surface. This is particularly true for electrically conductive materials such as metal and carbon fiber, making radar well-suited to the detection of aircraft and ships.

Radar absorbing material, containing resistive and sometimes magnetic substances, is used on military vehicles to reduce radar reflection. This is the radio equivalent of painting something a dark color so that it cannot be seen by the eye at night.

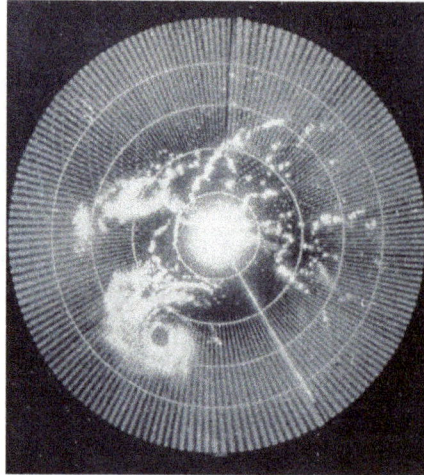

Brightness can indicate reflectivity as in this 1960 weather radar image (of Hurricane Abby). The radar's frequency, pulse form, polarization, signal processing, and antenna determine what it can observe.

Radar waves scatter in a variety of ways depending on the size (wavelength) of the radio wave and the shape of the target. If the wavelength is much shorter than the target's size, the wave will bounce off in a way similar to the way light is reflected by a mirror. If the wavelength is much longer than the size of the target, the target may not be visible because of poor reflection. Low-frequency radar technology is dependent on resonances for detection, but not identification, of targets. This is described by Rayleigh scattering, an effect that creates Earth's blue sky and red sunsets. When the two length scales are comparable, there may be resonances. Early radars used very long wavelengths that were larger than the targets and thus received a vague signal, where as some modern systems use shorter wavelengths (a few centimeters or less) that can image objects as small as a loaf of bread.

Short radio waves reflect from curves and corners in a way similar to glint from a rounded piece of glass. The most reflective targets for short wavelengths have 90° angles between the reflective surfaces. A corner reflector consists of three flat surfaces meeting like the inside corner of a box. The structure will reflect waves entering its opening directly back to the source. They are commonly used as radar reflectors to make otherwise difficult-to-detect objects easier to detect. Corner reflectors on boats, for example, make them more detectable to avoid collision or during a rescue. For similar reasons, objects intended to avoid detection will not have inside corners or surfaces and edges perpendicular to likely detection directions, which leads to "odd" looking stealth aircraft. These precautions do not completely eliminate reflection because of diffraction, especially at longer wavelengths. Half wavelength long wires or strips of conducting material, such as chaff, are very reflective but do not direct the scattered energy back toward the source. The extent to which an object reflects or scatters radio waves is called its radar cross section.

Radar Equation

The power P_r returning to the receiving antenna is given by the equation:

$$P_r = \frac{P_t G_t A_r \sigma F^4}{(4\pi)^2 R_t^2 R_r^2}$$

where

- P_t = transmitter power

- G_t = gain of the transmitting antenna

- A_r = effective aperture (area) of the receiving antenna; this can also be expressed as $\frac{G_r \lambda^2}{4\pi}$, where

    - λ^2 = transmitted wavelength

    - G_r = gain of receiving antenna

- σ = radar cross section, or scattering coefficient, of the target

- F = pattern propagation factor

- R_t = distance from the transmitter to the target

- R_r = distance from the target to the receiver.

In the common case where the transmitter and the receiver are at the same location, $R_t = R_r$ and the term $R_t^2 R_r^2$ can be replaced by R^4, where R is the range. This yields:

$$P_r = \frac{P_t G_t A_r \sigma F^4}{(4\pi)^2 R^4}$$

This shows that the received power declines as the fourth power of the range, which means that the received power from distant targets is relatively very small.

Additional filtering and pulse integration modifies the radar equation slightly for pulse-Doppler radar performance, which can be used to increase detection range and reduce transmit power.

The equation above with $F = 1$ is a simplification for transmission in a vacuum without interference. The propagation factor accounts for the effects of multipath and shadowing and depends on the details of the environment. In a real-world situation, pathloss effects should also be considered.

Doppler Effect

Frequency shift is caused by motion that changes the number of wavelengths between the reflector and the radar. That can degrade or enhance radar performance depending upon how that affects the detection process. As an example, Moving Target Indication can interact with Doppler to produce signal cancellation at certain radial velocities, which degrades performance.

Sea-based radar systems, semi-active radar homing, active radar homing, weather radar, military aircraft, and radar astronomy rely on the Doppler effect to enhance performance. This produces information about target velocity during the detection process. This also allows small objects to be detected in an environment containing much larger nearby slow moving objects.

Doppler shift depends upon whether the radar configuration is active or passive. Active radar transmits a signal that is reflected back to the receiver. Passive radar depends upon the object sending a signal to the receiver.

The Doppler frequency shift for active radar is as follows, where F_D is Doppler frequency, F_T is transmit frequency, V_R is radial velocity, and C is the speed of light:

$$F_D = 2 \times F_T \times \left(\frac{V_R}{C} \right).$$

Passive radar is applicable to electronic countermeasures and radio astronomy as follows:

$$F_D = F_T \times \left(\frac{V_R}{C} \right).$$

Only the radial component of the velocity is relevant. When the reflector is moving at right angle to the radar beam, it has no relative velocity. Vehicles and weather moving parallel to the radar beam produce the maximum Doppler frequency shift.

Doppler measurement is reliable only if the sampling rate exceeds the Nyquist frequency for the frequency shift produced by the radial motion. As an example, Doppler weather radar with a pulse rate of 2 kHz and transmit frequency of 1 GHz can reliably measure weather up to 150 m/s (340 mph) but cannot reliably determine radial velocity of aircraft moving 1,000 m/s (2,200 mph).

Polarization

In all electromagnetic radiation, the electric field is perpendicular to the direction of propagation, and the electric field direction is the polarization of the wave. For a transmitted radar signal, the polarization can be controlled to yield different effects. Radars use horizontal, vertical, linear, and circular polarization to detect different types of reflections. For example, circular polarization is used to minimize the interference caused by rain. Linear polarization returns usually indicate metal surfaces. Random polarization returns usually indicate a fractal surface, such as rocks or soil, and are used by navigation radars.

Limiting Factors

Beam Path and Range

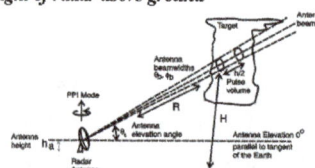

BEAM HEIGHT WITH DISTANCE (AGL)

$$H = \left(\sqrt{r^2 + (k_e a_e)^2 + 2r \, k_e a_e \sin(\theta_e)} \right) - k_e a_e + h_a$$

r : distance k_e :   4/3 (Standard refraction coefficient)

a_e :   Earth radius θ_e :   Elevation angle

h_a :   Height of radar above ground

Echo heights above ground

The radar beam would follow a linear path in vacuum, but it really follows a somewhat curved path in the atmosphere because of the variation of the refractive index of air, that is called the radar horizon. Even when the beam is emitted parallel to the ground, it will rise above it as the Earth curvature sinks below the horizon. Furthermore, the signal is attenuated by the medium it crosses, and the beam disperses.

The maximum range of a conventional radar can be limited by a number of factors:

- Line of sight, which depends on height above ground. This means without a direct line of sight the path of the beam is blocked.

- The maximum non-ambiguous range, which is determined by the pulse repetition frequency. The maximum non-ambiguous range is the distance the pulse could travel and return before the next pulse is emitted.

- Radar sensitivity and power of the return signal as computed in the radar equation. This includes factors such as environmental conditions and the size (or radar cross section) of the target.

Noise

Signal noise is an internal source of random variations in the signal, which is generated by all electronic components.

Reflected signals decline rapidly as distance increases, so noise introduces a radar range limitation. The noise floor and signal to noise ratio are two different measures of performance that affect range performance. Reflectors that are too far away produce too little signal to exceed the noise floor and cannot be detected. Detection requires a signal that exceeds the noise floor by at least the signal to noise ratio.

Noise typically appears as random variations superimposed on the desired echo signal received in the radar receiver. The lower the power of the desired signal, the more difficult it is to discern it from the noise. Noise figure is a measure of the noise produced by a receiver compared to an ideal receiver, and this needs to be minimized.

Shot noise is produced by electrons in transit across a discontinuity, which occurs in all detectors. Shot noise is the dominant source in most receivers. There will also be flicker noise caused by electron transit through amplification devices, which is reduced using heterodyne amplification. Another reason for heterodyne processing is that for fixed fractional bandwidth, the instantaneous bandwidth increases linearly in frequency. This allows improved range resolution. The one notable exception to heterodyne (downconversion) radar systems is ultra-wideband radar. Here a single cycle, or transient wave, is used similar to UWB communications.

Noise is also generated by external sources, most importantly the natural thermal radiation of the background surrounding the target of interest. In modern radar systems, the internal noise is typically about equal to or lower than the external noise. An exception is if the radar is aimed upwards at clear sky, where the scene is so "cold" that it generates very little thermal noise. The thermal noise is given by $k_B T B$, where T is temperature, B is bandwidth (post matched filter) and

k_B is Boltzmann's constant. There is an appealing intuitive interpretation of this relationship in a radar. Matched filtering allows the entire energy received from a target to be compressed into a single bin (be it a range, Doppler,elevation, or azimuth bin). On the surface it would appear that then within a fixed interval of time one could obtain perfect, error free, detection. To do this one simply compresses all energy into an infinitesimal time slice. What limits this approach in the real world is that, while time is arbitrarily divisible, current is not. The quantum of electrical energy is an electron, and so the best one can do is match filter all energy into a single electron. Since the electron is moving at a certain temperature (Plank spectrum) this noise source cannot be further eroded. We see then that radar, like all macro-scale entities, is profoundly impacted by quantum theory.

Noise is random and target signals are not. Signal processing can take advantage of this phenomenon to reduce the noise floor using two strategies. The kind of signal integration used with moving target indication can improve noise up to $\sqrt{2}$ for each stage. The signal can also be split among multiple filters for pulse-Doppler signal processing, which reduces the noise floor by the number of filters. These improvements depend upon coherence.

Interference

Radar systems must overcome unwanted signals in order to focus on the targets of interest. These unwanted signals may originate from internal and external sources, both passive and active. The ability of the radar system to overcome these unwanted signals defines its signal-to-noise ratio (SNR). SNR is defined as the ratio of the signal power to the noise power within the desired signal; it compares the level of a desired target signal to the level of background noise (atmospheric noise and noise generated within the receiver). The higher a system's SNR the better it is at discriminating actual targets from noise signals.

Clutter

Clutter refers to radio frequency (RF) echoes returned from targets which are uninteresting to the radar operators. Such targets include natural objects such as ground, sea, precipitation (such as rain, snow or hail), sand storms, animals (especially birds), atmospheric turbulence, and other atmospheric effects, such as ionosphere reflections, meteor trails, and Hail spike. Clutter may also be returned from man-made objects such as buildings and, intentionally, by radar countermeasures such as chaff.

Some clutter may also be caused by a long radar waveguide between the radar transceiver and the antenna. In a typical plan position indicator (PPI) radar with a rotating antenna, this will usually be seen as a "sun" or "sunburst" in the centre of the display as the receiver responds to echoes from dust particles and misguided RF in the waveguide. Adjusting the timing between when the transmitter sends a pulse and when the receiver stage is enabled will generally reduce the sunburst without affecting the accuracy of the range, since most sunburst is caused by a diffused transmit pulse reflected before it leaves the antenna. Clutter is considered a passive interference source, since it only appears in response to radar signals sent by the radar.

Clutter is detected and neutralized in several ways. Clutter tends to appear static between radar scans; on subsequent scan echoes, desirable targets will appear to move, and all stationary echoes can be eliminated. Sea clutter can be reduced by using horizontal polarization, while rain is reduced with circular polarization (note that meteorological radars wish for the opposite effect, and therefore use linear polarization to detect precipitation). Other methods attempt to increase the signal-to-clutter ratio.

Clutter moves with the wind or is stationary. Two common strategies to improve measure or performance in a clutter environment are:

- Moving target indication, which integrates successive pulses and

- Doppler processing, which uses filters to separate clutter from desirable signals.

The most effective clutter reduction technique is pulse-Doppler radar. Doppler separates clutter from aircraft and spacecraft using a frequency spectrum, so individual signals can be separated from multiple reflectors located in the same volume using velocity differences. This requires a coherent transmitter. Another technique uses a moving target indicator that subtracts the receive signal from two successive pulses using phase to reduce signals from slow moving objects. This can be adapted for systems that lack a coherent transmitter, such as time-domain pulse-amplitude radar.

Constant false alarm rate, a form of automatic gain control (AGC), is a method that relies on clutter returns far outnumbering echoes from targets of interest. The receiver's gain is automatically adjusted to maintain a constant level of overall visible clutter. While this does not help detect targets masked by stronger surrounding clutter, it does help to distinguish strong target sources. In the past, radar AGC was electronically controlled and affected the gain of the entire radar receiver. As radars evolved, AGC became computer-software controlled and affected the gain with greater granularity in specific detection cells.

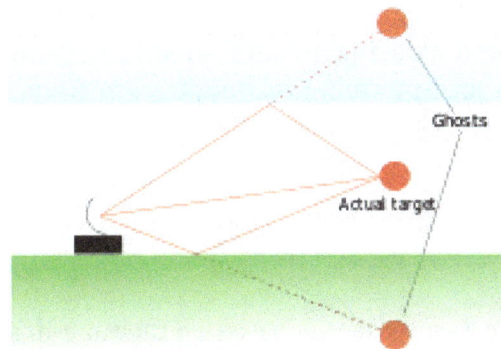

Radar multipath echoes from a target cause ghosts to appear.

Clutter may also originate from multipath echoes from valid targets caused by ground reflection, atmospheric ducting or ionospheric reflection/refraction (e.g., anomalous propagation). This clutter type is especially bothersome since it appears to move and behave like other normal (point) targets of interest. In a typical scenario, an aircraft echo is reflected from the ground below, appearing to the receiver as an identical target below the correct one. The radar may try to unify the targets, reporting the target at an incorrect height, or eliminating it on the basis of jitter or a physical impossibility. Terrain bounce jamming exploits this response by amplifying the radar signal and directing it downward. These problems can be overcome by incorporating a ground map of the radar's surroundings and eliminating all echoes which appear to originate below ground or above a certain height. Monopulse can be improved by altering the elevation algorithm used at low elevation. In newer air traffic control radar equipment, algorithms are used to identify the false targets by comparing the current pulse returns to those adjacent, as well as calculating return improbabilities.

Jamming

Radar jamming refers to radio frequency signals originating from sources outside the radar, transmitting in the radar's frequency and thereby masking targets of interest. Jamming may be intentional, as with an electronic warfare tactic, or unintentional, as with friendly forces operating equipment that transmits using the same frequency range. Jamming is considered an active interference source, since it is initiated by elements outside the radar and in general unrelated to the radar signals.

Jamming is problematic to radar since the jamming signal only needs to travel one way (from the jammer to the radar receiver) whereas the radar echoes travel two ways (radar-target-radar) and are therefore significantly reduced in power by the time they return to the radar receiver. Jammers therefore can be much less powerful than their jammed radars and still effectively mask targets along the line of sight from the jammer to the radar (*mainlobe jamming*). Jammers have an added effect of affecting radars along other lines of sight through the radar receiver's sidelobes (*sidelobe jamming*).

Mainlobe jamming can generally only be reduced by narrowing the mainlobe solid angle and cannot fully be eliminated when directly facing a jammer which uses the same frequency and polarization as the radar. Sidelobe jamming can be overcome by reducing receiving sidelobes in the radar antenna design and by using an omnidirectional antenna to detect and disregard non-mainlobe signals. Other anti-jamming techniques are frequency hopping and polarization.

Radar Signal Processing

Transit Time

One way to obtain a distance measurement is based on the time-of-flight: transmit a short pulse of radio signal (electromagnetic radiation) and measure the time it takes for the reflection to return. The distance is one-half the product of the round trip time (because the signal has to travel to the target and then back to the receiver) and the speed of the signal. Since radio waves travel at the speed of light, accurate distance measurement requires high-performance electronics. In most cases, the receiver does not detect the return while the signal is being transmitted. Through the use of a duplexer, the radar switches between transmitting and receiving at a predetermined rate. A similar effect imposes a maximum range as well. In order to maximize range, longer times between pulses should be used, referred to as a pulse repetition time, or its reciprocal, pulse repetition frequency.

Pulse radar: The round-trip time for the radar pulse to get to the target and return is measured. The distance is proportional to this time.

These two effects tend to be at odds with each other, and it is not easy to combine both good short range and good long range in a single radar. This is because the short pulses needed for a good minimum range broadcast have less total energy, making the returns much smaller and the target harder to detect. This could be offset by using more pulses, but this would shorten the maximum range. So each radar uses a particular type of signal. Long-range radars tend to use long pulses with long delays between them, and short range radars use smaller pulses with less time between them. As electronics have improved many radars now can change their pulse repetition frequency, thereby changing their range. The newest radars fire two pulses during one cell, one for short range (10 km / 6 miles) and a separate signal for longer ranges (100 km /60 miles).

The distance resolution and the characteristics of the received signal as compared to noise depends on the shape of the pulse. The pulse is often modulated to achieve better performance using a technique known as pulse compression.

Distance may also be measured as a function of time. The radar mile is the amount of time it takes for a radar pulse to travel one nautical mile, reflect off a target, and return to the radar antenna. Since a nautical mile is defined as 1,852 meters, then dividing this distance by the speed of light (299,792,458 meters per second), and then multiplying the result by 2 yields a result of 12.36 microseconds in duration.

Frequency Modulation

Another form of distance measuring radar is based on frequency modulation. Frequency comparison between two signals is considerably more accurate, even with older electronics, than timing the signal. By measuring the frequency of the returned signal and comparing that with the original, the difference can be easily measured.

This technique can be used in continuous wave radar and is often found in aircraft radar altimeters. In these systems a "carrier" radar signal is frequency modulated in a predictable way, typically varying up and down with a sine wave or sawtooth pattern at audio frequencies. The signal is then sent out from one antenna and received on another, typically located on the bottom of the aircraft, and the signal can be continuously compared using a simple *beat frequency* modulator that produces an audio frequency tone from the returned signal and a portion of the transmitted signal.

Since the signal frequency is changing, by the time the signal returns to the aircraft the transmit frequency has changed. The amount of frequency shift is used to measure distance.

The modulation index riding on the receive signal is proportional to the time delay between the radar and the reflector. The amount of that frequency shift becomes greater with greater time delay. The measure of the amount of frequency shift is directly proportional to the distance traveled. That distance can be displayed on an instrument, and it may also be available via the transponder. This signal processing is similar to that used in speed detecting Doppler radar. Example systems using this approach are AZUSA, MISTRAM, and UDOP.

A further advantage is that the radar can operate effectively at relatively low frequencies. This was important in the early development of this type when high frequency signal generation was difficult or expensive.

Terrestrial radar uses low-power FM signals that cover a larger frequency range. The multiple reflections are analyzed mathematically for pattern changes with multiple passes creating a computerized synthetic image. Doppler effects are used which allows slow moving objects to be detected as well as largely eliminating "noise" from the surfaces of bodies of water.

Speed Measurement

Speed is the change in distance to an object with respect to time. Thus the existing system for measuring distance, combined with a memory capacity to see where the target last was, is enough to measure speed. At one time the memory consisted of a user making grease pencil marks on the radar screen and then calculating the speed using a slide rule. Modern radar systems perform the equivalent operation faster and more accurately using computers.

If the transmitter's output is coherent (phase synchronized), there is another effect that can be used to make almost instant speed measurements (no memory is required), known as the Doppler effect. Most modern radar systems use this principle into Doppler radar and pulse-Doppler radar systems (weather radar, military radar, etc...). The Doppler effect is only able to determine the relative speed of the target along the line of sight from the radar to the target. Any component of target velocity perpendicular to the line of sight cannot be determined by using the Doppler effect alone, but it can be determined by tracking the target's azimuth over time.

It is possible to make a Doppler radar without any pulsing, known as a continuous-wave radar (CW radar), by sending out a very pure signal of a known frequency. CW radar is ideal for determining the radial component of a target's velocity. CW radar is typically used by traffic enforcement to measure vehicle speed quickly and accurately where range is not important.

When using a pulsed radar, the variation between the phase of successive returns gives the distance the target has moved between pulses, and thus its speed can be calculated. Other mathematical developments in radar signal processing include time-frequency analysis (Weyl Heisenberg or wavelet), as well as the chirplet transform which makes use of the change of frequency of returns from moving targets ("chirp").

Pulse-doppler Signal Processing

Pulse-Doppler signal processing includes frequency filtering in the detection process. The space between each transmit pulse is divided into range cells or range gates. Each cell is filtered independently much like the process used by a spectrum analyzer to produce the display showing different frequencies. Each different distance produces a different spectrum. These spectra are used to perform the detection process. This is required to achieve acceptable performance in hostile environments involving weather, terrain, and electronic countermeasures.

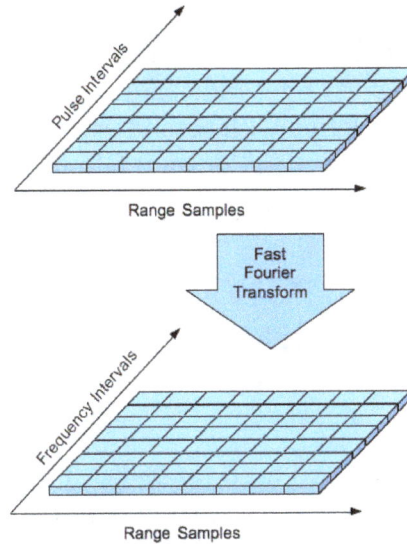

Pulse-Doppler signal processing. The *Range Sample* axis represents individual samples taken in between each transmit pulse. The *Range Interval* axis represents each successive transmit pulse interval during which samples are taken. The Fast Fourier Transform process converts time-domain samples into frequency domain spectra. This is sometimes called the *bed of nails*.

The primary purpose is to measure both the amplitude and frequency of the aggregate reflected signal from multiple distances. This is used with weather radar to measure radial wind velocity and precipitation rate in each different volume of air. This is linked with computing systems to produce a real-time electronic weather map. Aircraft safety depends upon continuous access to accurate weather radar information that is used to prevent injuries and accidents. Weather radar uses a low PRF. Coherency requirements are not as strict as those for military systems because individual signals ordinarily do not need to be separated. Less sophisticated filtering is required, and range ambiguity processing is not normally needed with weather radar in comparison with military radar intended to track air vehicles.

The alternate purpose is "look-down/shoot-down" capability required to improve military air combat survivability. Pulse-Doppler is also used for ground based surveillance radar required to defend personnel and vehicles. Pulse-Doppler signal processing increases the maximum detection distance using less radiation in close proximity to aircraft pilots, shipboard personnel, infantry, and artillery. Reflections from terrain, water, and weather produce signals much larger than aircraft and missiles, which allows fast moving vehicles to hide using nap-of-the-earth flying techniques and stealth technology to avoid detection until an attack vehicle is too close to destroy. Pulse-Doppler signal processing incorporates more sophisticated electronic filtering that safely eliminates this kind of weakness. This requires the use of medium pulse-repetition frequency with phase coherent hardware that has a large dynamic range. Military applications require medium PRF which prevents range from being determined directly, and range ambiguity resolution processing is required to identify the true range of all reflected signals. Radial movement is usually linked with Doppler frequency to produce a lock signal that cannot be produced by radar jamming signals. Pulse-Doppler signal processing also produces audible signals that can be used for threat identification.

Reduction of Interference Effects

Signal processing is employed in radar systems to reduce the radar interference effects. Signal processing techniques include moving target indication, Pulse-Doppler signal processing, moving target detection processors, correlation with secondary surveillance radar targets, space-time adaptive processing, and track-before-detect. Constant false alarm rate and digital terrain model processing are also used in clutter environments.

Plot and Track Extraction

A Track algorithm is a radar performance enhancement strategy. Tracking algorithms provide the ability to predict future position of multiple moving objects based on the history of the individual positions being reported by sensor systems.

Historical information is accumulated and used to predict future position for use with air traffic control, threat estimation, combat system doctrine, gun aiming, and missile guidance. Position data is accumulated radar sensors over the span of a few minutes.

There are four common track algorithms.

- Nearest Neighbor

- Probabilistic Data Association

- Multiple Hypothesis Tracking

- Interactive Multiple Model (IMM)

Radar video returns from aircraft can be subjected to a plot extraction process whereby spurious and interfering signals are discarded. A sequence of target returns can be monitored through a device known as a plot extractor.

The non-relevant real time returns can be removed from the displayed information and a single plot displayed. In some radar systems, or alternatively in the command and control system to which the radar is connected, a radar tracker is used to associate the sequence of plots belonging to individual targets and estimate the targets' headings and speeds.

Engineering

A radar's components are:

- A transmitter that generates the radio signal with an oscillator such as a klystron or a magnetron and controls its duration by a modulator.

- A waveguide that links the transmitter and the antenna.

- A duplexer that serves as a switch between the antenna and the transmitter or the receiver for the signal when the antenna is used in both situations.

- A receiver. Knowing the shape of the desired received signal (a pulse), an optimal receiver can be designed using a matched filter.

- A display processor to produce signals for human readable output devices.

- An electronic section that controls all those devices and the antenna to perform the radar scan ordered by software.

- A link to end user devices and displays.

Components of a Radar/Composantes d'un radar

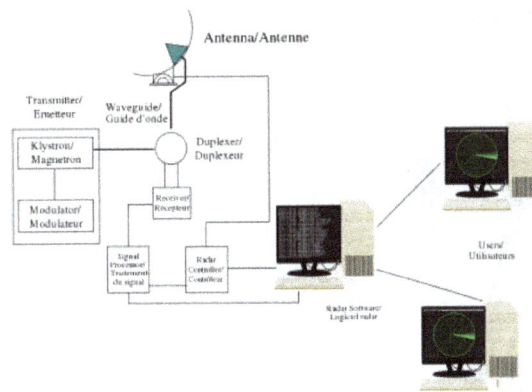

Radar components

Antenna Design

Radio signals broadcast from a single antenna will spread out in all directions, and likewise a single antenna will receive signals equally from all directions. This leaves the radar with the problem of deciding where the target object is located.

Early systems tended to use omnidirectional broadcast antennas, with directional receiver antennas which were pointed in various directions. For instance, the first system to be deployed, Chain Home, used two straight antennas at right angles for reception, each on a different display. The maximum return would be detected with an antenna at right angles to the target, and a minimum with the antenna pointed directly at it (end on). The operator could determine the direction to a target by rotating the antenna so one display showed a maximum while the other showed a minimum. One serious limitation with this type of solution is that the broadcast is sent out in all directions, so the amount of energy in the direction being examined is a small part of that transmitted. To get a reasonable amount of power on the "target", the transmitting aerial should also be directional.

Parabolic Reflector

More modern systems use a steerable parabolic "dish" to create a tight broadcast beam, typically using the same dish as the receiver. Such systems often combine two radar frequencies in the same antenna in order to allow automatic steering, or *radar lock*.

Parabolic reflectors can be either symmetric parabolas or spoiled parabolas: Symmetric parabolic antennas produce a narrow "pencil" beam in both the X and Y dimensions and consequently have a higher gain. The NEXRAD Pulse-Doppler weather radar uses a symmetric antenna to perform detailed volumetric scans of the atmosphere. Spoiled parabolic antennas produce a narrow beam in one dimension and a relatively wide beam in the other. This feature is useful if target detection over a wide range of angles is more important than target location in three dimensions. Most 2D surveillance radars use a spoiled parabolic antenna with a narrow azimuthal beamwidth and wide vertical beamwidth. This beam configuration allows the radar operator to detect an aircraft at a specific azimuth but at an indeterminate height. Conversely, so-called "nodder" height finding radars use a dish with a narrow vertical beamwidth and wide azimuthal beamwidth to detect an aircraft at a specific height but with low azimuthal precision.

Surveillance radar antenna

Types of Scan

- Primary Scan: A scanning technique where the main antenna aerial is moved to produce a scanning beam, examples include circular scan, sector scan, etc.

- Secondary Scan: A scanning technique where the antenna feed is moved to produce a scanning beam, examples include conical scan, unidirectional sector scan, lobe switching, etc.

- Palmer Scan: A scanning technique that produces a scanning beam by moving the main antenna and its feed. A Palmer Scan is a combination of a Primary Scan and a Secondary Scan.

- Conical scanning: The radar beam is rotated in a small circle around the "boresight" axis, which is pointed at the target.

Slotted Waveguide

Applied similarly to the parabolic reflector, the slotted waveguide is moved mechanically to scan and is particularly suitable for non-tracking surface scan systems, where the vertical pattern may remain constant. Owing to its lower cost and less wind exposure, shipboard, airport surface, and harbour surveillance radars now use this approach in preference to a parabolic antenna.

Slotted waveguide antenna

Phased Array

Another method of steering is used in a phased array radar.

Phased array: Not all radar antennas must rotate to scan the sky.

Phased array antennas are composed of evenly spaced similar antenna elements, such as aerials or rows of slotted waveguide. Each antenna element or group of antenna elements incorporates a discrete phase shift that produces a phase gradient across the array. For example, array elements producing a 5 degree phase shift for each wavelength across the array face will produce a beam pointed 5 degree away from the centerline perpendicular to the array face. Signals traveling along that beam will be reinforced. Signals offset from that beam will be canceled. The amount of reinforcement is antenna gain. The amount of cancellation is side-lobe suppression.

Phased array radars have been in use since the earliest years of radar in World War II (Mammut radar), but electronic device limitations led to poor performance. Phased array radars were originally used for missile defense. They are the heart of the ship-borne Aegis Combat System and the Patriot Missile System. The massive redundancy asso-ciated with having a large number of array elements increases reliability at the expense of gradual performance degradation that occurs as individual phase elements fail.

Phased array antenna can be built to conform to specific shapes, like missiles, infantry support vehicles, ships, and aircraft.

As the price of electronics has fallen, phased array radars have become more common. Almost all modern military radar systems are based on phased arrays, where the small additional cost is offset by the improved reliability of a system with no moving parts. Traditional moving-antenna designs are still widely used in roles where cost is a significant factor such as air traffic surveillance, weather radars and similar systems.

Phased array radars are valued for use in aircraft since they can track multiple targets. The first aircraft to use a phased array radar was the B-1B Lancer. The first fighter aircraft to use phased array radar was the Mikoyan MiG-31. The MiG-31M's SBI-16 Zaslon Passive electronically scanned array radar was considered to be the world's most powerful fighter radar, until the AN/APG-77 Active electronically scanned array was introduced on the Lockheed Martin F-22 Raptor.

Phased-array interferometry or aperture synthesis techniques, using an array of separate dishes that are phased into a single effective aperture, are not typical for radar applications, although they are widely used in radio astronomy. Because of the thinned array curse, such multiple aperture arrays, when used in transmitters, result in narrow beams at the expense of reducing the total power transmitted to the target. In principle, such techniques could increase spatial resolution, but the lower power means that this is generally not effective.

Aperture synthesis by post-processing motion data from a single moving source, on the other hand, is widely used in space and airborne radar systems .

Frequency Bands

The traditional band names originated as code-names during World War II and are still in military and aviation use throughout the world. They have been adopted in the United States by the Institute of Electrical and Electronics Engineers and internationally by the International Telecommunication Union. Most countries have additional regulations to control which parts of each band are available for civilian or military use.

Other users of the radio spectrum, such as the broadcasting and electronic countermeasures industries, have replaced the traditional military designations with their own systems.

Radar frequency bands			
Band name	**Frequency range**	**Wavelength range**	**Notes**
HF	3–30 MHz	10–100 m	Coastal radar systems, over-the-horizon radar (OTH) radars; 'high frequency'
VHF	30–300 MHz	1–10 m	Very long range, ground penetrating; 'very high frequency'
P	< 300 MHz	> 1 m	'P' for 'previous', applied retrospectively to early radar systems; essentially HF + VHF
UHF	300–1000 MHz	0.3–1 m	Very long range (e.g. ballistic missile early warning), ground penetrating, foliage penetrating; 'ultra high frequency'
L	1–2 GHz	15–30 cm	Long range air traffic control and surveillance; 'L' for 'long'

S	2–4 GHz	7.5–15 cm	Moderate range surveillance, Terminal air traffic control, long-range weather, marine radar; 'S' for 'short'
C	4–8 GHz	3.75–7.5 cm	Satellite transponders; a compromise (hence 'C') between X and S bands; weather; long range tracking
X	8–12 GHz	2.5–3.75 cm	Missile guidance, marine radar, weather, medium-resolution mapping and ground surveillance; in the United States the narrow range 10.525 GHz ±25 MHz is used for airport radar; short range tracking. Named X band because the frequency was a secret during WW2.
K_u	12–18 GHz	1.67–2.5 cm	High-resolution, also used for satellite transponders, frequency under K band (hence 'u')
K	18–24 GHz	1.11–1.67 cm	From German *kurz*, meaning 'short'; limited use due to absorption by water vapour, so K_u and K_a were used instead for surveillance. K-band is used for detecting clouds by meteorologists, and by police for detecting speeding motorists. K-band radar guns operate at 24.150 ± 0.100 GHz.
K_a	24–40 GHz	0.75–1.11 cm	Mapping, short range, airport surveillance; frequency just above K band (hence 'a') Photo radar, used to trigger cameras which take pictures of license plates of cars running red lights, operates at 34.300 ± 0.100 GHz.
mm	40–300 GHz	1.0–7.5 mm	Millimetre band, subdivided as below. The frequency ranges depend on waveguide size. Multiple letters are assigned to these bands by different groups. These are from Baytron, a now defunct company that made test equipment.
V	40–75 GHz	4.0–7.5 mm	Very strongly absorbed by atmospheric oxygen, which resonates at 60 GHz.
W	75–110 GHz	2.7–4.0 mm	Used as a visual sensor for experimental autonomous vehicles, high-resolution meteorological observation, and imaging.

Radar Modulators

Modulators act to provide the waveform of the RF-pulse. There are two different radar modulator designs:

- High voltage switch for non-coherent keyed power-oscillators These modulators consist of a high voltage pulse generator formed from a high voltage supply, a pulse forming network, and a high voltage switch such as a thyratron. They generate short pulses of power to feed, e.g., the magnetron, a special type of vacuum tube that converts DC (usually pulsed) into microwaves. This technology is known as pulsed power. In this way, the transmitted pulse of RF radiation is kept to a defined and usually very short duration.

- Hybrid mixers, fed by a waveform generator and an exciter for a complex but coherent waveform. This waveform can be generated by low power/low-voltage input signals. In this case the radar transmitter must be a power-amplifier, e.g., a klystron tube or a solid state transmitter. In this way, the transmitted pulse is intrapulse-modulated and the radar receiver must use pulse compression techniques.

Radar Coolant

Coherent microwave amplifiers operating above 1,000 watts microwave output, like traveling wave tubes and klystrons, require liquid coolant. The electron beam must contain 5 to 10 times more power than the microwave output, which can produce enough heat to generate plasma. This plasma flows from the collector toward the cathode. The same magnetic focusing that guides the electron beam forces the plasma into the path of the electron beam but flowing in the opposite direction. This introduces FM modulation which degrades Doppler performance. To prevent this, liquid coolant with minimum pressure and flow rate is required, and deionized water is normally used in most high power surface radar systems that utilize Doppler processing.

Coolanol (silicate ester) was used in several military radars in the 1970s. However, it is hygroscopic, leading to hydrolysis and formation of highly flammable alcohol. The loss of a U.S. Navy aircraft in 1978 was attributed to a silicate ester fire. Coolanol is also expensive and toxic. The U.S. Navy has instituted a program named Pollution Prevention (P2) to eliminate or reduce the volume and toxicity of waste, air emissions, and effluent discharges. Because of this, Coolanol is used less often today.

Regulations

Radar (also: *RADAR*) is defined by *article 1.100* of the International Telecommunication Union´s (ITU) ITU Radio Regulations (RR) as:

A radiodetermination system based on the comparison of reference signals with radio signals reflected, or retransmitted, from the position to be determined. Each *radiodetermination system* shall be classified by the *radiocommunication service* in which it operates permanently or temporarily. Typical radar utilizations are primary radar and secondary radar, these might operate in the radiolocation service or the radiolocation-satellite service.

Radar Imaging

Imaging radar is an application of radar which is used to create two-dimensional images, typically of landscapes. Imaging radar provides its light to illuminate an area on the ground and take a picture at radio wavelengths. It uses an antenna and digital computer storage to record its images. In a radar image, one can see only the energy that was reflected back towards the radar antenna. The radar moves along a flight path and the area illuminated by the radar, or footprint, is moved along the surface in a swath, building the image as it does so.

An SAR radar image acquired by the SIR-C/X-SAR radar on board the Space Shuttle Endeavour shows the Teide volcano. The city of Santa Cruz de Tenerife is visible as the purple and white area on the lower right edge of the island. Lava flows at the summit crater appear in shades of green and brown, while vegetation zones appear as areas of purple, green and yellow on the volcano's flanks.

Digital radar images are composed of many dots. Each pixel in the radar image represents the radar backscatter for that area on the ground: brighter areas represent high backscatter, darker areas represents low backscatter.

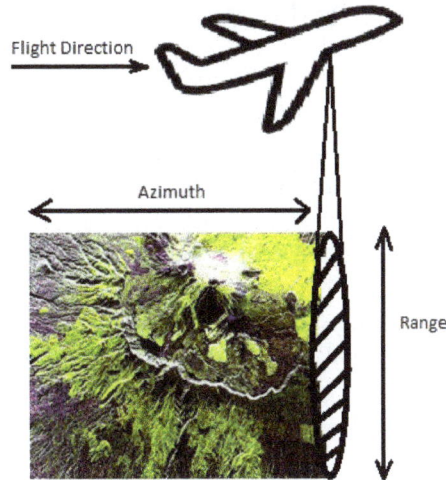

Building up a radar image using the motion of the platform

The traditional application of radar is to display the position and motion of typically highly reflective objects (such as aircraft or ships) by sending out an radiowave signal, and then detecting the direction and delay of the reflected signal. Imaging radar on the other hand attempts to form an image of one object (e.g. a landscape) by furthermore registering the intensity of the reflected signal to determine the amount of scattering (cf. Light scattering). The registered electromagnetic scattering is then mapped onto a two-dimensional plane, with points with a higher reflectivity getting assigned usually a brighter color, thus creating an image.

Several techniques have evolved to do this. Generally they take advantage of the Doppler effect caused by the rotation or other motion of the object and by the changing view of the object brought about by the relative motion between the object and the back-scatter that is perceived by the radar of the object (typically, a plane) flying over the earth. Through recent improvements of the techniques, radar imaging is getting more accurate. Imaging radar has been used to map the Earth, other planets, asteroids, other celestial objects and to categorize targets for military systems.

Imaging Radar

An imaging radar is a kind of radar equipment which can be used for imaging. A typical radar technology includes emitting radio waves, receiving their reflection, and using this information to generate data. For an imaging radar, the returning waves are used to create an image. When the radio waves reflect off objects, this will make some changes in the radio waves and can provide data about the objects, including how far the waves traveled and what kind of objects they encountered. Using the acquired data, a computer can create a 3-D or 2-D image of the target.

Imaging radar has several advantages. It can operate in the presence of obstacles that obscure the target, and can penetrate ground (sand), water, or walls.

Applications

Applications include: surface topography & crustal change; land use monitoring, agricultural monitoring, ice patrol, environmental monitoring;weather radar- storm monitoring, wind shear warning;medical microwave tomography; through wall radar imaging; 3-D measurements, etc.

Through Wall Radar Imaging

Wall parameter estimation uses Utra Wide-Band radar systems. The handle M-sequence UWB radar with horn and circular antennas was used for data gathering and supporting the scanning method.

3-D Measurements

3-D measurements are supplied by amplitude-modulated laser radars—Erim sensor and Perceptron sensor. In terms of speed and reliability for median-range operations, 3-D measurements have superior performance.

Techniques and Methods

Current radar imaging techniques rely mainly on synthetic aperture radar (SAR) and inverse synthetic aperture radar (ISAR) imaging. Emerging technology utilizes monopulse radar 3-D imaging.

Real Aperture Radar

Real aperture radar(RAR) is a form of radar that transmits a narrow angle beam of pulse radio wave in the range direction at tight angles to the flight direction and receives the backscattering from the targets which will be transformed to a radar image from the received signals.

Usually the reflected pulse will be arranged in the order of return time from the targets, which corresponds to the range direction scanning.

The resolution in the range direction depends on the pulse width. The resolution in the azimuth direction is identical to the multiplication of beam width and the distance to a target.

AVTIS Radar

The AVTIS radar is a 94 GHz real aperture 3D imaging radar. It uses Frequency-Modulated Continuous-Wave(FMCW) modulation and employs a mechanically scanned monostatic with sub-metre range resolution.

Laser Radar

Laser radar is a remote sensing technology that measures distance by illuminating a target with a laser and analyzing the reflected light.

Laser radar is used for multi-dimensional imaging and information gathering. In all information gathering modes, lasers that transmit in the eye-safe region are required as well as sensitive receivers at these wavelengths.

3-D imaging requires the capacity to measure the range to the first scatter within every pixel. Hence, an array of range counters is needed. A monolithic approach to an array of range counters is being developed. This technology must be coupled with highly sensitive detectors of eye-safe wavelengths.

To measure Doppler information requires a different type of detection scheme than is used for spatial imaging. The returned laser energy must be mixed with a local oscillator in a heterodyne system to allow extraction of the Doppler shift.

3-D, Multi-Wave and Multi-Band, Imaging Radar

3-D, Multi-wave and Multi-band, Imaging radar works in one of the two modes - as an arbitrary frequency(KHz-MHz), arbitrary wave radar, and as a C-band analog and digital mode radar. The system architecture of 3-D, Multi-wave and Multi-band, Imaging radar is shown in the figure.

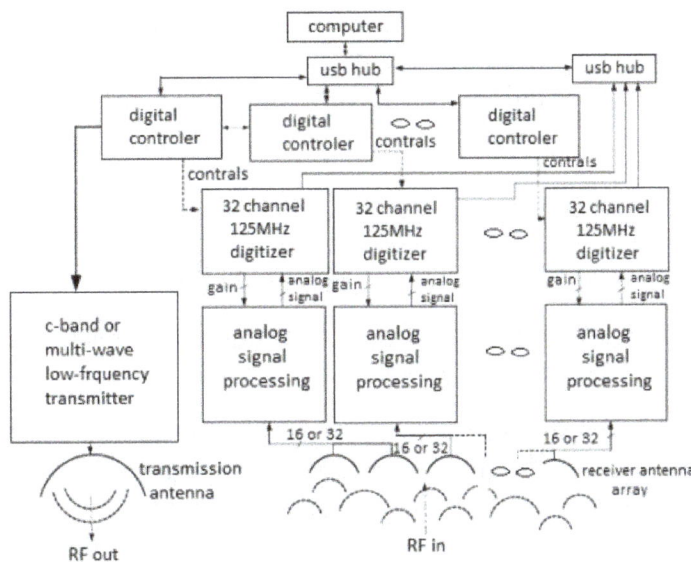

3-D, Multi-wave, Multi-band, Imaging Radar Architecture

Synthetic Aperture Radar (SAR)

Synthetic-aperture radar (SAR) is a form of radar which moves a real aperture or antenna through a series of positions along the objects to provide distinctive long-term coherent-signal variations. This can be used to obtain higher resolution.

SARs produce a two-dimensional (2-D) image. One dimension in the image is called range and is a measure of the "line-of-sight" distance from the radar to the object. Range is determined by measuring the time from transmission of a pulse to receiving the echo from a target. Also, range resolution is determined by the transmitted pulse width.The other dimension is called azimuth and is perpendicular to range. The ability of SAR of producing relatively fine azimuth resolution makes it different from other radars.To obtain fine azimuth resolution, a physically large antenna is needed to focus the transmitted and received energy into a sharp beam. The sharpness of the beam defines the azimuth resolution. An airborne radar could collect data while flying this distance and process the data as if it came from a physically long antenna. The distance the aircraft flies in synthesizing the antenna is known as the synthetic aperture. A narrow synthetic beamwidth results from the relatively long synthetic aperture, which gets finer resolution than a smaller physical antenna.

Inverse Aperture Radar (ISAR)

Inverse synthetic aperture radar (ISAR) is another kind of SAR system which can produce high-resolution on two- and three-dimensional images.

An ISAR system consists of a stationary radar antenna and a target scene that is undergoing some motion. ISAR is theoretically equivalent to SAR in that high-azimuth resolution is achieved via relative motion between the sensor and object, yet the ISAR moving target scene is usually made up of non cooperative objects.

Algorithms with more complex schemes for motion error correction are needed for ISAR imaging than those needed in SAR. ISAR technology uses the movement of the target rather than the emitter to make the synthetic aperture. ISAR radars are commonly used on vessels or aircraft and can provide a radar image of sufficient quality for target recognition. The ISAR image is often adequate to discriminate between various missiles, military aircraft, and civilian aircraft.

Disadvantages of ISAR

1. The ISAR imaging cannot obtain the real azimuth of the target.

2. There sometimes exists a reverse image. For example, the image formed of a boat when it rolls forwards and backwards in the ocean.

3. The ISAR image is the 2-D projection image of the target on the Range-Doppler plane which is perpendicular to the rotating axis. When the Range-Doppler plane and the coordinate plane are different, the ISAR image can not reflect the real shape of the target. Thus, the ISAR imaging can not obtain the real shape information of the target in most situations.

Monopulse Radar 3-D Imaging Technique

Monopulse radar 3-D imaging technique uses 1-D range image and monopulse angle measurement to get the real coordinates of each scatterer. Using this technique, the image doesn't vary with the change of the target's movement. Monopulse radar 3-D imaging utilizes the ISAR techniques to separate scatterers in the Doppler domain and perform monopulse angle measurement.

Monopulse radar 3-D imaging can obtain the 3 views of 3-D objects by using any two of the three parameters obtained from the azimuth difference beam, elevation difference beam and range measurement, which means the views of front, top and side can be azimuth-elevation, azimuth-range and elevation-range, respectively.

Monopulse imaging generally adapts to near-range targets, and the image obtained by monopulse radar 3-D imaging is the physical image which is consistent with the real size of the object.

Radar Navigation

Marine and aviation radar systems can provide very useful navigation information in a variety of situations. When a vessel is within radar range of land or special radar aids to navigation, the navigator can take distances and angular bearings to charted objects and use these to establish arcs of position and lines of position on a chart. A fix consisting of only radar information is called a radar fix.

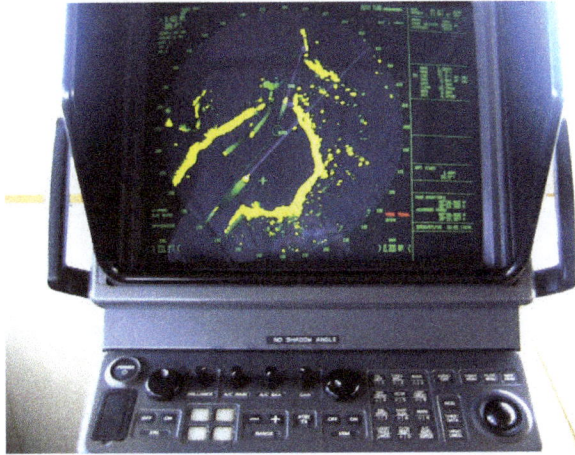

Radar ranges and bearings can be very useful for navigation.

Some types of radar fixes include the relatively self-explanatory methods of "range and bearing to a single object," "two or more bearings," "tangent bearings," and "two or more ranges."

Parallel indexing is a technique defined by William Burger in the 1957 book *The Radar Observer's Handbook*. This technique involves creating a line on the screen that is parallel to the ship's course, but offset to the left or right by some distance. This parallel line allows the navigator to maintain a given distance away from hazards.

Some techniques have been developed for special situations. One, known as the "contour method," involves marking a transparent plastic template on the radar screen and moving it to the chart to fix a position.

Another special technique, known as the Franklin Continuous Radar Plot Technique, involves drawing the path a radar object should follow on the radar display if the ship stays on its planned course. During the transit, the navigator can check that the ship is on track by checking that the pip lies on the drawn line.

After completing the plotting radar technique, the image from the radar can either be displayed, captured or recorded to a computer monitor using a frame grabber.

References

- Watson, Raymond C., Jr. (2009-11-25). Radar Origins Worldwide: History of Its Evolution in 13 Nations Through World War II. Trafford Publishing. ISBN 978-1-4269-2111-7.

- Aftanas, Michal (2010). Through-Wall Imaging With UWB Radar System (PDF). Berlin: LAP LAMBERT Academic Publishing. p. 132. ISBN 3838391764.

- Maloney, Elbert S. (December 2003). Chapman Piloting and Seamanship (64th ed.). New York, NY: Hearst Communications Inc. ISBN 1-58816-089-0.

- Bowditch, Nathaniel (2002). The American Practical Navigator. Bethesda, MD: National Imagery and Mapping Agency. ISBN 0-939837-54-4.

- Cutler, Thomas J. (December 2003). Dutton's Nautical Navigation (15th ed.). Annapolis, MD: Naval Institute Press. ISBN 978-1-55750-248-3.

- "Discover the Benefits of Radar Imaging « Earth Imaging Journal: Remote Sensing, Satellite Images, Satellite Imagery". eijournal.com. Retrieved 2015-11-13.

Permissions

Index

A

Active Sonar, 58, 61-68, 71-76, 198

Actuators, 1-8, 10, 12, 14, 16, 18, 20, 22, 24, 26, 28, 30, 32, 34, 36, 38, 40, 42, 44, 46, 48, 50, 52, 54, 56, 58, 60, 62, 64, 66, 68, 70, 72, 74, 76, 78, 80, 82, 84, 86, 88, 90, 92, 94, 96, 98, 100, 102, 104, 106, 108, 110-192, 194, 196, 198, 200, 202, 204, 206, 208, 210, 212, 214, 216, 218, 220, 222, 224, 228, 230, 232, 234, 236, 238, 240, 242, 244, 246, 248, 250, 252, 254

Area Monitoring, 53

Axial Plunger Motors, 123

B

Ball Screw, 7, 126, 130, 133-134, 136, 168, 181

Bathymetric Mapping, 75

Bioreceptors, 90-91

Biosensor, 4, 89-96, 98-99

Biosensor System, 89

Biotransducer, 89, 92

Blood Glucose Meters, 100-101

Blood Glucose Monitoring, 100, 102-104

C

Comb Drive, 6, 138

D

Data Formats, 10

Data Logger, 9-14

Demodulation, 22, 43, 208

Distributed Sensor Network, 57

Doppler Effect, 25, 63, 231, 233, 240, 249

E

Echo Sounding, 9, 64, 73, 77-78, 80

Electric Actuators, 149

Electro-mechanical Actuators, 128

Electroactive Polymers, 152, 172, 177, 192

Erdas Imagine, 203-204, 217-220

F

Fluid Power Actuators, 151

G

Geodetic, 12, 26, 44, 198, 216

Gerotor Motors, 123

Global Positioning System, 9, 20, 24, 30, 109, 207

Glucose Sensing Bio-implants, 103

H

Health Care Monitoring, 53

Helical Band Actuator, 132, 143

Hydraulic Cylinder, 5-6, 112-113, 115-121, 127, 162, 164

Hydraulic Load Cell, 108

Hydraulic Motor, 112, 121-122, 124-125

Hydraulic Press, 153, 164-166, 168

Hydrographic Echo Sounding, 80

I

Illumination, 199-200, 211, 231

Industrial Metal Detectors, 15, 21

Industrial Monitoring, 54

Instrumentation Protocols, 10

Intercept Sonar, 72

Inverted Roller Screw, 184

Ion Channel Switch, 93-94

J

Jackscrew, 127, 153, 166-169

L

Laser Radar, 207, 250

Level Sensor, 80, 82, 87-88

Lidar, 196-197, 204-217, 223-225, 227

Linear Actuator, 125-126, 128-131, 133, 136-137, 139, 143, 186

Load Cell, 9, 104-109

M

Magnetoresistive, 87

Magnetostrictive Actuators, 189

Mems Magnetic Actuator, 186

Metal Detector, 14-21

N

Nanoparticles, 99

O

Optical Interface, 85

Ozone Biosensors, 98

P

Parametric Sonar, 75
Passive Sonar, 58, 61, 65-66, 68-71, 198
Photoelectric Sensor, 22-23
Piezoelectric Actuators, 7, 128
Piezoelectric Load Cell, 108
Piston Rod Construction, 117
Pneumatic Actuator, 5-6, 110-111
Pneumatic Cylinder, 153, 158-161
Pneumatic Gripper, 111-112
Pneumatic Load Cell, 108
Pneumatic Motor, 153-155
Polarization, 62, 174, 178, 213, 232, 234, 237-238
Precise Monitoring, 46
Pulse Induction, 18

R

Rack And Pinion, 5-7, 126, 136-138, 140, 143, 151
Radar, 58, 61, 65-66, 68, 70, 81, 85-87, 195-197, 201-202, 204-205, 207, 218-219, 221, 223-224, 226-254
Radar Equation, 232-233, 235
Radar Imaging, 248-250, 254
Radar Navigation, 253
Radar Signal, 227, 231, 234, 238-240
Radar Signal Processing, 238, 240
Radial Piston Motors, 124
Remote Sensing Software, 203, 220
Retraction Force Difference, 113
Rigid Belt Actuator, 132, 142-143
Rigid Chain Actuator, 132, 139-141, 143
Roller Screw, 7, 126-127, 153, 180-182, 184-186
Rotary Actuator, 112, 149-151
Rotating Paddle, 81

S

Satellite Frequencies, 42
Scattering, 67, 194, 206, 214-215, 232-233, 249
Security Screening, 14, 21
Sensing Modes, 23
Sensors, 1-109, 112, 114, 116, 118, 120, 122, 124, 126, 128, 130, 132, 134, 136, 138, 140, 142, 144, 146, 148, 150, 152, 154, 156, 158, 160, 162, 164, 166, 168, 170, 172, 174, 176, 178-180, 182, 184, 186, 188, 190, 192, 194, 196-198, 200, 202, 204, 206-208, 210-212, 214-216, 218, 220, 222, 224, 228, 230, 232, 234, 236, 238, 240, 242, 244, 246, 248, 250, 252, 254
Side Loading, 118, 160
Sonar, 9, 58, 60-78, 198
Sound Propagation, 58, 66-67

T

Telescoping Linear Actuator, 131
Terrset, 220-221
Timekeeping, 47
Transponder, 26, 64, 240

U

User-satellite Geometry, 31

V

Vibrating Point, 81
Vortex Generator, 147

W

Wall Radar Imaging, 250
Water Remote Sensing, 193-194
Wireless Sensor Network, 52-53, 56-57
Wiring Colors, 108

www.ingramcontent.com/pod-product-compliance
Lightning Source LLC
Chambersburg PA
CBHW061305190326
41458CB00011B/3773